T0290037

MATHEMATICS RESEARCH DEVELOPMENTS

PERSPECTIVES IN APPLIED MATHEMATICS

Mathematics Research Developments

Additional books in this series can be found on Nova's website under the Series tab.

Additional E-books in this series can be found on Nova's website under the E-books tab.

MATHEMATICS RESEARCH DEVELOPMENTS

PERSPECTIVES IN APPLIED MATHEMATICS

JORDAN I. CAMPBELL
EDITOR

Nova Science Publishers, Inc.
New York

For permission to use material from this book please contact us:
Telephone 631-231-7269; Fax 631-231-8175
Web Site: http://www.novapublishers.com

LIBRARY OF CONGRESS CATALOGING-IN-PUBLICATION DATA

Perspectives in applied mathematics / editor: Jordan I. Campbell.
p. cm.
Includes bibliographical references and index.
ISBN 978-1-61122-796-3 (hardcover)
1. Applied mathematics. I. Campbell, Jordan I.
QA3.P45 2010
510--dc22

2010043924

Published by Nova Science Publishers, Inc. † New York

CONTENTS

PREFACE

This book presents and discusses the role mathematics plays with regards to various social, business and biological issues. Topics discussed include the coalition formation process; substitutable and perishable inventory with partial backlogging; contractual stability and competitive equilibrium in a pure exchange economy; a heterogeneous foraging swarm model inspired by honeybee behavior and a cautionary tale of caterpillars and selectional interference. (Imprint: Nova Press)

There are many models that try to explain the formation of social networks, for example the coalitions of agents. Such models are useful to understand the alliances formed in wars, economical conflicts, political parties, etc. Most of the models use two-body interactions to simulate the relationship between agents, supposing that a pair interaction can't be affected by the rest of the agents in the system. However, in this work we show that such model is not good enough to explain all real phenomena that occur in coalition formation. Thus, as happens in nuclear physics, many body interactions have to be considered in social models. Particularly, in Chapter 1 we present a study of the effects of three-body interactions in the process of coalition formation, modifying a spin glass model of bimodal propensities in order to include a particular three-body Hamiltonian that reproduces the main features of the required interactions. We apply the model to a simplificated scenario of the Iraq war. For the calculation of the interaction parameters between agents, we propose the usage of the renormalization group theory to include all internal degrees of freedom of the social system.

Chapter 2 presents a two commodity stochastic perishable inventory system under continuous review. The maximum storage capacity for the i-th item is fixed as S_i $(i = 1,2)$. It is assumed that demand for the i-th commodity is of unit size and independent of the other commodity. Demand time points form a Poisson process with parameter λ_i, $i=1,2$. The life time of each item of i-th commodity is exponential with parameter γ_i, $i=1,2$. The reorder level is fixed as s_i for the i-th commodity $(i=1,2)$ and the ordering policy is to place order for $Q_i(=S_i-s_i)$ items of the i-th commodity $(i=1,2)$ when both the inventory levels are less than or equal to their respective reorder levels. The lead time for replenishment is assumed to be exponential with parameter μ. The two commodities are assumed to be substitutable. That is, if the inventory level of one commodity reaches zero, then any demand for this commodity will be satisfied by the other commodity. If substitute is not available, then this demand is backlogged up to a certain level N_i, $(i=1,2)$ for the i-th commodity. Whenever the inventory level reaches N_i, $(i=1,2)$, an order for N_i items is placed and replenished instantaneously. For this model, the limiting probability distribution for the joint inventory levels is computed. Various operational characteristics and expression for long run total expected cost rate are derived.

Chapter 3 looks at finding a formula to determine the best applicant applying to a 3 memeber committee. Management and computer ability are both being considered.

Chapter 4 contains a game-theoretical analysis of the so-called weak totally contractual allocations, similar to that introduced by V.L.Makarov [4] in order to describe stable outcomes of some quite natural recontracting processes in pure exchange economies. Detailed presentation of Makarov's original settings can be found in [5]. Below, we analyze a slightly strengthened version of contractual blocking, introduced in this paper. Namely, in what follows, no additional restrictions to the stopping rule of the breaking procedure is posed besides the feasibility of the final contractual system (hence, no minimality condition, applied in [4-6], and [8,9]). At the same time, like in [4-6], any such a final contractual system supposed to be an improvement for each member of blocking coalition (not just at least one of the minimal final system, like it appears to be in [8,9]).

Chapter 5 is organized as follows: In Section 2, the authors formulate a model of one-person game and prove a general theorem showing that an optimal stopping time problem with continuous time, discount function and random independent horizon is equivalent to an optimal stopping problem with a new discount factor. This theorem is a generalization to continuous time of the theorem presented by Samuel-Cahn (1996). They obtain the solution of our model and discuss its properties. Moreover, the influence of stochastic order and hazard rate order on the optimal mean reward is considered. In Section 3, a generalization of our one person game with random horizon to a multi-person game with random horizon is analyzed. Examples are presented in Section 4.

Chapter 6 provides a strategic dynamic analysis of a theoretical scenario where *verifiable* information is transmitted between an informed sender and an uninformed but rational decision maker in a multidimensional space. The information provided by the sender is assumed to be encoded in *multifunctions*. The authors show that each one of these multifunctions induces a preference relation on the decision maker. These induced preferences do not generally coincide with those defined in a complete information environment. They provide sufficient conditions for the utility functions induced by the information sender to be continuous. In addition, they use homotopy theory to illustrate how the information encoded by the sender in the multifunctions can be modified through time in a continuous way so as to induce *any* a priori assigned preference relation on the decision maker.

Inspired by the organized behaviors of honeybee swarms, an individual based mathematical model is proposed in Chapter 7 for the heterogeneous swarm. The heterogeneous swarm is assumed to consist of two different kinds of individuals, namely, the scouts and the normal agents, with respect to their sensing abilities. Besides, a short-distance-bounded-attraction function was proposed to describe the attraction among individuals.

Firstly the heterogeneous swarm model is identified and the swarm cohesion is proved, and the analytical bound on the swarm size is provided. Secondly, the foraging properties of the heterogeneous swarm in multimodal Gaussian environment are studied, and conditions for collective convergence to more favorable regions are provided. Thirdly, simulations were carried out and the priority of proposed short-distance-bounded-attraction function was demonstrated in complex environment. Simulation results show that the heterogeneous swarm model provides a feasible framework for multi-robot navigation applications.

Evolutionary theory yields many important insights into why organisms are the way they are. However, for any given problem, there is significant uncertainty about what level of abstraction is appropriate. With such issues in mind, an evolutionary scenario concerning the size and development time of a particular species of caterpillar, the tobacco hornworm will be discussed in

Chapter 8. A larva of this species grows approximately exponentially; once it reaches a critical size various hormones result in a time delay before pupation during which it continues to grow. Thus, the relationship between genotype and phenotype is non-trivial. In laboratory experiments by Nijhout and in computer simulations, one can selectively breed caterpillars based on size and development time and investigate the resulting evolutionary dynamics. Both laboratory experiments and computer simulations yield striking complexity. Selecting for size and time simultaneously yields unexpected interference, resulting in caterpillars that do not necessarily have the selected-for properties. This interference is due to constraints on the set of possible phenotypes imposed by the caterpillar's fundamental development process. The important lesson here is that when building evolutionary models of even simple phenomena, one must be distrustful of intuition, and consider (1) how the details of the development process of individual organisms affect the ability of a population to explore the space of possible phenotypes, and (2) how selection criteria in combination can interact in unexpected ways.

In: Perspectives in Applied Mathematics ISBN 978-1-61122-796-3
Editor: Jordan I. Campbell, pp. 1-26

Chapter 1

MANY-BODY EFFECTS IN THE COALITION FORMATION PROCESS

F. Samaniego-Steta [1], ***Gerardo G. Naumis*** [2,3,*]
and M. del Castillo-Mussot [4]

[1] Departamento de Física-Química, Instituto de Física,
Universidad Nacional Autónoma de México,
Apartado Postal 20-364, 01000 México, Distrito Federal, México
[2] Facultad de Ciencias, Universidad Autónoma del Estado de Morelos,
Av. Universidad 10001, 62210 Cuernavaca, Morelos, México
[3] Departamento de Física-Matematica, Universidad Iberoamericana,
Prolongación Paseo de la Reforma 880, Col. Lomas de Santa Fe,
01210 México, Distrito Federal, México
[*] On sabbatical leave
[4] Departamento de Estado Sólido, Instituto de Física,
Universidad Nacional Autónoma de México, Apartado Postal 20-364,
01000 México, Distrito Federal, México

Abstract

There are many models that try to explain the formation of social networks, for example the coalitions of agents. Such models are useful to understand the alliances formed in wars, economical conflicts, political parties, etc. Most of the models use two-body interactions to simulate the relationship between agents, supposing that a pair interaction can't be affected by the rest of the agents in the system. However, in this work we show that such model is not good enough to explain all real phenomena that occur in coalition formation. Thus, as happens in nuclear physics, many body interactions have to be considered in social models. Particularly, we present a study of the effects of three-body interactions in the process of coalition formation, modifying a spin glass model of bimodal propensities in order to include a particular three-body Hamiltonian that reproduces the main features of the required interactions. We apply the model to a simplificated scenario of the Iraq war. For the calculation of the interaction parameters between agents, we propose the usage of the renormalization group theory to include all internal degrees of freedom of the social system.

1. Introduction

How does social conflict appear? The motivation to model conflicts between social agents is supported by the idea that there has to be some kind of algorithm that leaders (chiefs of state, the president of a political party, or any other persons who take responsability for the decisions of a group), use in order to make the decisions that will start a fight between two or more social agents. However, all the decisions taken in real conflicts are different and therefore it seems it is very difficult to obtain any regulariy. Can we think about an algorithm which captures the main ingredients of this social phenomena to explore the possibilites in a situation of antagonism or even war?

The attempts to model the origins and behaviour of the social networks are growing in the scientific community. However there are many issues that have not been considered and can not be omitted to model these systems correctly. For example, the nature of the interactions between agents is very important in social relationships, so in order to improve the models created until now it is necessary to consider aspects that has not been treated in previous works. Such aspects is the consideration of many-body interactions. We have many empirical evidence that lead us to include this type of interactions between social agents. For example, we can observe them in the relationships between the triangle formed by a boy, his girlfriend and his mother in law. The relation that the girl have with her boyfriend will affect directly the relation between her mother and the boy. In international affairs this fact also occurs. For example Mexico is very dependent on the USA, and any international relation that Mexico could establish with any country of the Middle East will be affected by the relationship between the USA and the Middle East. In physics these many-body interactions (specially the three-body ones) have been studied for several phenomena. For example, in nuclear or high-energy physics these interactions emerge naturally [1][2]. In contrast, the long range coulombic and gravitational forces are two-body interactions. In many other branches of physics these interactions are employed, for example in the modelation of polymers, as well as in atomic and molecular physics to simulate torsion, bending and general bonds [3][4][5].

There is a previous attempt to consider three-body interactions in social science [6], and in 2005, Lambiotte and Ausloos [7] presented the importance of more than two collaborations in the social network. This helped to think about networks where the basic connections were not binary, but triangles instead.

Almost all the social systems have important many-body interactions, and the formation of coalitions as a form of aggregation is not an exception. For example in political parties, parliaments, fusions of companies, and in warfare coalition forming is important. To model and study this social phenomenon we need to understand primarly this type of relationships. These systems have been studied using concepts from the theory of spin glasses, i.e., disordered materials that exhibit a high magnetic frustration due to competing interactions, where the interactions between agents are represented by the coupling parameter of the corresponding spin model. The tendency of two social agents either to be in conflict or to cooperate is going to be simulated by their type of interaction. In principle, any social system in which polarization arises, can be studied applying the models that will be presented below.

The first attempt to model the formation of coalitions was made by Axelrod and Bennet (model AB), in the seminal paper "A Landscape theory of Aggregation"[8]. They used a minimum energy principle to construct a model of landscapes of aggregation. This model was applied to the seventeen European nations that were involved in the Second World War (WW II) [8], and to nine computer companies that were in conflict to set the standards for the UNIX operating system [9]. Florian and Galam improved the previous model to study the fragmentation of former Yugoslavia[10].

However, these models were made under the supposition that each agent can only interact with one of their neighbours in the network bilaterally. In other words the interactions are in pairs and there are not any considerations of many-body interactions. The main objective of this work is to show that such interactions have important quantitative and qualitative consequences and they can not be omitted in the social models. Thus, we propose an extension of one of the existing models (Galam model), were we consider the simplest one of these interactions: the three-body case. However, even if we only work with the simplest of these interactions, this study is good enough to show the effects that they have over the process of coalition formation.

Another aspect that should be considered in these models are the internal degrees of freedom of each social agent, in order to make a more accurate representation of them and to reduce the fuzziness that emerges from the calculation of the existing parameters. For instance, the foreign policy of a country or its position in an international conflict is often stated to be determined in great manner by the internal social or class struggles.

Following the spirit of applying spin glass concepts to the formation of alliances, we present here a model that includes the three-body interactions. Furthermore, we explicitely showed its performance in an important and recent geopolitical event such as the 2003 invasion of Iraq. We compare our model with the available ones in literature were three body effects have not been considered before. Afterwards we present some ideas to include the internal degrees of freedom of social agents employing the renormalization group theory, which originally was developed for physical systems.

2. Coalition Formation Models

2.1. The Axelrod-Bennett Model (AB Model)

In 1993 Robert Axelrod and Scott Bennett used the principle of minimum energy to build a model based on the landscapes of aggregation. They published "A Landscape Theory of Aggregation"[8]; where aggregation was understood as the organization of the elements of a system in patterns where the similar ones tend to be together and the the others to move apart.

This model adresses the problem of alignment between a group of n countries that seek to form coalitions. The main contribution of this work is the consideration of pairwise propensities p_{ij} between the agents i and j. This parameter can be seen as how willing are two agents to be part of any of the coalitions. The propensities are assumed to be symmetric, i.e. $p_{ij} = p_{ji}$.

A configuration of the system is a partition of the n agents, i.e. each actor enters in one of the coalitions. In this case they dealt with the simplest case and only two coalitions are possible, therefore they worked with a system of bimodal coalitions.

For a given configuration X Axelrod and Bennett defined the distance between the agents i and j as $d_{ij}(X)$, where $d_{ij}(X) = 0$ when the actors are in the same alliance, and 1 if they are in a different one.

Following the introduction of distances $d_{ij}(X)$'s and the pairwise propensities, a new quantity called frustration is introduced ($F_i(X)$), which depends on the given configuration X and it's calculated for every actor i in the network as follows:

$$F_i(X) = \sum_{j=1}^{n} s_j p_{ij} d_{ij}(X) \tag{1}$$

where s_i is the size or strength of the i-th agent, which can be quantified taking into account some considerations such as demographic, industrial, military, or a combination of these factors, depending on the weight that we are assigning on each of the agents. The sum is taken over all the actors considering that $p_{ii} = 0$. This frustration $F_i(X)$ can be understood as a measure of the satisfaction that the agents have when they establish contact with another one, or in other words, it gives a quantification of how great is the conflict between agents.

The addition of the frustrations for all the agents is defined as the energy of the system,

$$E(X) = \sum_{i} s_i F_i(X), \tag{2}$$

where the sum runs over the $n(n-1)/2$ distinct pairs (i, j). Each social agent has two choices of coalition, therefore there are by symmetry $2^n/2$ distinct sets of alliances. The equation (2) is the central expression for the AB model.

This model postulates that the actual configuration is the one which minimizes the energy. To accomplish this, the pairwise propensities are defined as $p_{ij} > 0$ when the agents i and j tend to be aligned and $p_{ij} < 0$ when they are not, so with this new parameter the model assure that the most probable configuration is that one which minimizes the energy.

Axelrod and Bennett made the following statements about their model:

1. About the size of the agents they comment: "Size plays a role because having a proper relationship with a large country is more important than having a proper relationship with a small country" [8, pp. 215].

2. The physical concept of frustration [11] is expressed in the model with "For example, if there are three nations that mutually dislike each other (such as Israel, Syria and Iraq), then any possible bipolar configuration will leave someone frustrated" [8, pp. 217].

3. Predictions can be made: “Landscape theory begins with sizes and pair-wise propensities that are used to calculate the energy of each possible configuration and then uses the resulting landscape to make predictions about the dynamics of the system” [8, pp. 217].

The sizes of the nations s_i are very important in the model AB, in order to obtain a more precise configuration, thus it is necessary to determine their values correctly.

This model was applied to the alliances formed at the Second World War II [8], and to the coalitions formed while setting the standards for the UNIX operating system [9], and the results were encouraging.

2.2. Another Perspective on the AB Model

Serge Galam showed that in bimodal coalitions, the AB model is equivalent to a finite, non-frustrated spin glass at zero temperature zero, or more specifically, to the Ising ferromagnetic model at $T = 0$ [12]. To prove the point, he proposed two coalitions A and B. After this he associated a variable η_i to each agent, where $\eta_i = 1$ if the actor is part of coalition A, and $\eta_i = -1$ if it is part of B. From the symmetry of the system all the agents that form the coalition A can be in the other coalition by switching all the actors in the coalition B to the A and viceversa.

Given a pair of actors (i, j), their respective alignment is determined by the product $\eta_i\eta_j$, which is equal to 1 if both actors belong to the same coalition, and is equal to -1 if they are in a different one. Using this new set of variables $\{\eta_i\}$, the distance between the actors i and j can be reformulated as follows,

$$d_{ij}(X) = \frac{1}{2}\left(1 - \eta_i(X)\eta_j(X)\right), \tag{3}$$

and the energy of the system becomes,

$$E(X) = E_0 - \frac{1}{2}\sum_{j>i}^{n} J_{ij}\eta_i(X)\eta_j(X), \tag{4}$$

where

$$J_{ij} \equiv s_i s_j p_{ij}, \tag{5}$$

with $J_{ii} = 0$ and

$$E_0 = \frac{1}{2}\sum_{j>i}^{n} J_{ij}. \tag{6}$$

E_0 is a constant which depends on the size of the social agents and on the bilateral propensities. However this constant is independent of the coalition distribution, so it does not have any effect on the dynamic of the alignments between actors. Therefore, we can eliminate E_0 it from the previous equation (4). Thus, one can obtain the following Hamiltonian,

$$H = -\frac{1}{2}\sum_{j>i}^{n} J_{ij}\eta_i\eta_j, \tag{7}$$

which has to be minimized according to $\{\eta_i\}$ knowing previously the set $\{J_{ij}\}$. Equation (7) corresponds to the Hamiltonian of the Ising model for competing interactions [11]. The new variables J_{ij}'s represent the coupling variables between spins at this physical model. In this case, when $J_{ij} > 0$ there is cooperation between the actors, meanwhile for $J_{ij} < 0$ the agents are in conflict.

After this change of variables, Galam states in his work [12] that the stable configuration is the one which minimizes the energy. Thus, the AB model is indeed at temperature $T = 0$. Otherwise, when $T \neq 0$ the free-energy has to be minimized. In practice, for a finite system the theory can tell which coalitions are possible and how many of them exist. When several coalitions have the same energy, it is not possible to predict which one will be the actual one.

2.3. The Galam Model (G Model)

Serge Galam proposed a modified model with a group of N actors and two coalitions A and B[12] adding an external alignment pressure. The bilateral propensities between the agents i and j are defined as $J_{i,j} \equiv J'_{i,j}$, which as in the AB model are calculated considering the historical, economical and cultural factors. These propensities are $J'_{i,j} > 0$ when there is cooperation between the agents, $J'_{i,j} < 0$ if there is conflict, and $J'_{i,j} = 0$ if there is ignorance or neutrality. Symmetry between them is maintained and the defined propensities $J'_{i,j}$ are local, since they don't take into account any global organization or network.

In this model Galam introduces the concept of natural belonging to either of the two competing coalitions (A or B), assigning a variable ε_i to each agent i, where $\varepsilon_i = 1$ if the actor should be in the coalition A, $\varepsilon_i = -1$ for the coalition B, and $\varepsilon_i = 0$ if there is no coalition in which the i agent should be. These natural belongings are induced by cultural, political and historical interests.

Within a global framework (with two world-level coalitions), the benefit $C_{i,j}$ obtained by each actor through the exchanges that this actor makes with the rest is always positive since sharing resources, information or weapons is always profitable. However the propensity of a pair of actors (i, j) to cooperate, to be in conflict or to ignore each other is given by $J_{i,j} = \varepsilon_i \varepsilon_j C_{i,j}$, which can be positive, negative or zero respectively.

Taking into account the local and global exchanges the model suggests a total propensity,

$$J_{i,j} = J'_{i,j} + \varepsilon_i \varepsilon_j C_{i,j}, \tag{8}$$

between two actors i and j with $C_{i,j} > 0$. Here we consider that each agent has connections with the rest, and thus the network has all the possible bonds.

As in the AB model, the actual belonging of an actor i to any of the two coalitions is determined by the spin variables η_i, where $\eta_i = 1$ if the agent belongs to the group A, $\eta_i = -1$ for the group B.

The military or economical pressures exerted on the alignments of some agents are considered by a magnetic field variable β_i. In case of favoring coalition A $\beta_i = 1$, for the B $\beta_i = -1$. In the case of no pressure then $\beta_i = 0$. The magnitude of these external factors is considered through an external positve field b_i, which accounts for the size and the importance of the agent affected. The sets $\{\varepsilon_i\}$ and $\{\beta_i\}$ are independent.

The Hamiltonian of this system is given by

$$H^{(2)} \equiv -\frac{1}{2}\sum_{i>j}^{N} J_{i,j}\eta_i\eta_j - \sum_{i}^{N} \beta_i b_i \eta_i, \tag{9}$$

which is basically the Ising Hamiltonian in his ground state for spin glasses with a local external magnetic field $h_i = \beta_i b_i$.

In this model several scenarios are considered, depending each of them on the values of the established parameters [12, 13]. In terms of these parameters, the possible scenarios will be [13]: local coalitions ($|J'_{ij}| \approx C_{ij}$), global cold war scenario ($|J'_{ij}| \ll C_{ij}$) or unique leader, in which $|J'_{ij}| \ll C_{ij}$ for the powerful leader coalition and $|J'_{ij}| \approx C_{ij}$ for the others actors that interact locally and with the leader.

This model was used qualitatively to explain the stability of the alliances during the Cold War, considering the NATO and the Warsaw pact [12], the recent West European construction as well as the stability of China [12]. In a later work [10] Galam and Florian employed this model to study the dismembering of former Yugoslavia (1991-1992), and they found that the number of optimal coalitions was greater than two.

2.4. Failure of Two-Body Models in the Iraq War

However, in all of these previous models, many-body effects were neglected although they are very important. To illustrate this point, we analyze a simplification of the Iraq war [14](started in 2003), using the two-body models shown before.

Our simplification is based on some ideas taken from a Renormalization Group procedure which will be stated later on and roughly applied to social structures. We imagine that only four actors were important to prove our point, so we considered: Iraq (Q), a Muslim Coalition (M), the United States of America (U) and Israel (I). All the other countries of the world were not considered cause they lead essentially to the same results, so we decided that this application should stay as simple as it could be. We considered a Muslim coalition since they share a common religion, similar cultural background, history and strong economic ties. It is worthwhile mentioning that the Muslim Coalition is just a label to denote all Muslim countries that did not send troops to Iraq (practically all). Notice that in the 1991 Gulf War against Iraq some Muslim countries sent troops to attack Iraq, because initially Iraq took over Kuwait (which is also Muslim). However, in the 2003 Iraq invasion there was not such strong division beacuse the causes of the war were different and did not involve harming or invading any other Muslim country.

In table 1, we show the natural belongings of each nation according to their history in previous conflicts (such as the 1991 Gulf war) and their economical and political interests.

According to history, the strongest degree of natural belonging in this conflict is between U and I which leads to a big parameter C_{U-I}. To obtain these parameters we considered that the United States (U) tried to invade Iraq in the 1991 Gulf war, and they have supported the government of Israel on several occasions.

Table 1. Natural belongings for each *i* country.

Country	ε_i
Q	-1
I	$+1$
M	-1
U	$+1$

Table 2. Interaction parameters J_{ij} between the countr *i* and the *j*.

	Q	I	M	U
Q	0	C_E	C	C_E
I	C_E	0	C_E	C_F
M	C	C_E	0	C
U	C_E	C_F	C	0

The labels of the interaction parameters between the nations are shown in table 2. C_F (strong friendship), C_E (strong enemy) and C (weak relationship) were chosen in order to maintain a hierarchy (following the G model): $C_F > C_E > C \gg |J'_{ij}|$. Using these parameters and the main expression of the G model (9), we found using numerical simulations, the ground state of the Hamiltonian[1] $H^{(2)}$. For almost all paramteres, I enters into the war allied with U in a coalition A, and Q takes the same side as M at the coalition B (Figure 2.1, Figure 2.2 a)). The corresponding ground state has an energy $E_0^{(2)} = -(C_F + 3C_E + 2C)$. The only relevant change in the solution, is achieved by assuming that a strong leader, like U, has a huge magnetic field, $b \gg 1$, that enforces M to enter into a coalition with U.

[1]The previous models had only taked into account the pairwise interactions. In order to state a difference between the two-body expressions from the three-body ones, we've inserted a superindex (2) at the first ones, and a (3) on the second ones.

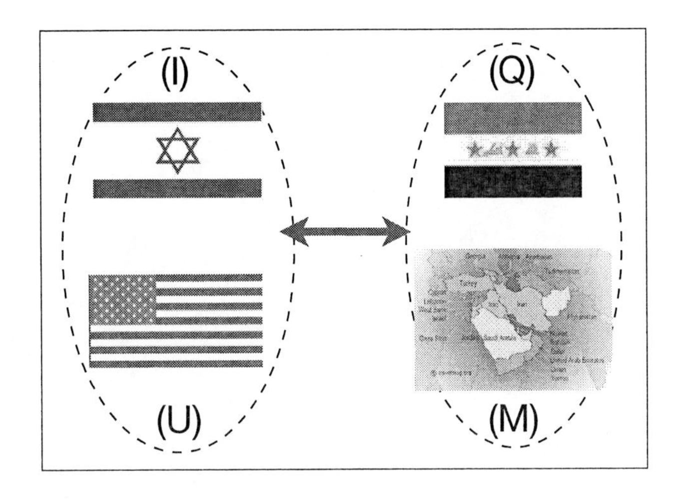

Figure 1. Coalitions obtained through the G model for the Iraq War example. The letters represen each social agent considered: (I) for Israel, (U) for USA, (Q) for Iraq, and (M) for the Muslim Coalition. The Muslim Coalition denote all the Muslim countries that did not send troops to Iraq in the 2003 invasion. For illustration purposes we only show in this map some of them in the neighborhood of Iraq together with other non Muslim countries.

After this calculations, we can see that the results presented by this model are not the same as the real situation in this conflict, militarily the Muslims and Israel stayed as neutral countries and they not entered into the war (Figure 2.2 b)). Within the $H^{(2)}$ model, to account for the possibility of neutrality of I and M, we are going to compare the minimal energy of the previous network with the case in which M and I are disconnected. In a scenario of strong enemies and friendships, assuming all $C_{ij} \gg J'$, the solution with neutral M and I has energy $E^{(2)}_{0,U-Q} = -C_E$. Since $C_F > C_E$ is clear that $E^{(2)}_0 < E^{(2)}_{0,U-Q}$, so the energy minimum is to keep all countries fighting. However, since the solution presented by the G model was not observed in the real situation, the main question is:

Why this result was not observed?

The main reason is the three-body interaction, and the associated damages due to war. If I goes into the coalition with U, the reaction of M will be very strong against U. Thus, the interaction between U and M depends also on I. In terms of the original idea of "distances between countries" of the AB model[8],

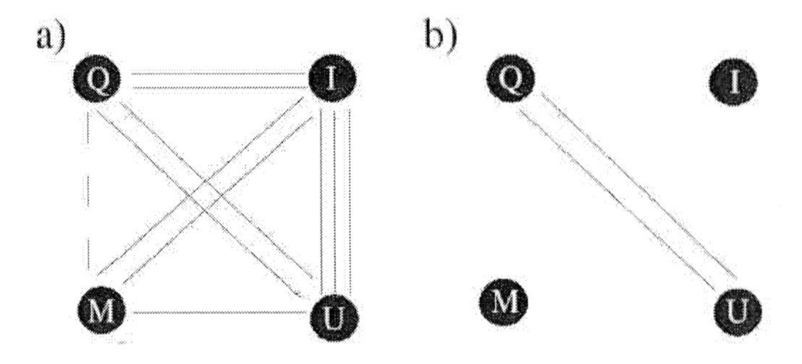

Figure 2. a) Simplificated network about the Iraq War example. A strong friendship (C_F) is represented by three solid lines, a strong enemy(C_E) by two lines, and the enemies (friends) with a weak interaction between them(C) by one solid line (dashed). b) Same network but with M and I as neutral contries, thus they are disconnected.

if the distance X between two of them is reduced, the other distance is increased! Thus our two-body interactions failed to give the real military coalition of Figure 2.2 b).

3. Modelling Three-Body Social Interactions

As we have seen from the previous example, previous models need to be improved in order to describe more precisely those phenomena which depend on the three-body interactions of their actors. To take into account these interactions we are going to consider triangles were the agents are placed at each vertex. They will interact through the bonds represented by the sides of the polygon. Notice that in general, a complex network contains many triangles. Each of them is a possible source for a three-body interaction (Figure 3.1).

3.1. Construction of the Three-Body Interaction Model

The proposal of a new model [14] has to reproduce the previous results that we have observed from the former models, since these two-body models still capture the basic ingredientes of the coalition formation process. Therefore we

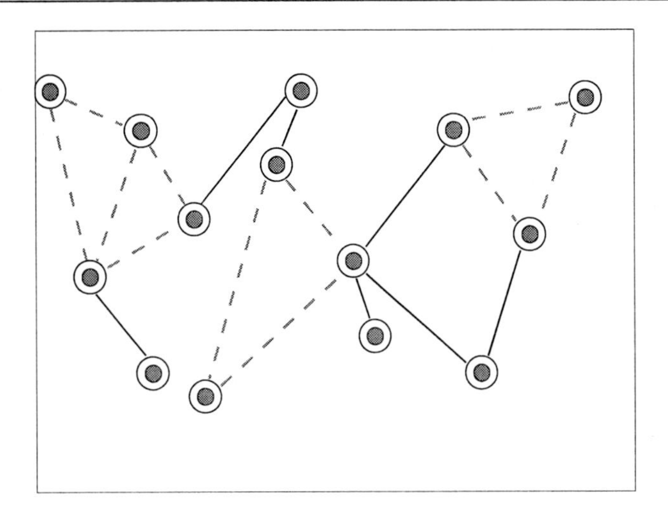

Figure 3. A network were each node represents a social agent. The triangles formed between three agents are presented with dashed lines.

proposed a modification adding an extra Hamiltonian $H^{(3)}$ to the original one $H^{(2)}$,

$$H = H^{(2)} + \alpha H^{(3)}, \tag{10}$$

where α is a linear parameter that measures the magnitude of the three-body effects. The most simple form of the perturbation is,

$$H^{(3)} = \sum_{i,j,k}^{N} \frac{t_{ijk}}{3} \eta_i \eta_j \eta_k, \tag{11}$$

with a coupling parameter t_{ijk} for each triangle of actors i,j and k that occurs in the lattice. The intensity of these connections will vary as in the former models, and they are given by,

$$t_{ijk} \equiv \gamma_{ijk} J_{ij} J_{jk} J_{ki}, \tag{12}$$

where γ_{ijk} is the magnitude of the conflict or damage associated with a three-body interaction between the three actors i, j and k. The bilateral propensities correspond to those given by the G model (8). However, this simple Hamiltonian (11) doesn't consider some important situations that can occur in triangles

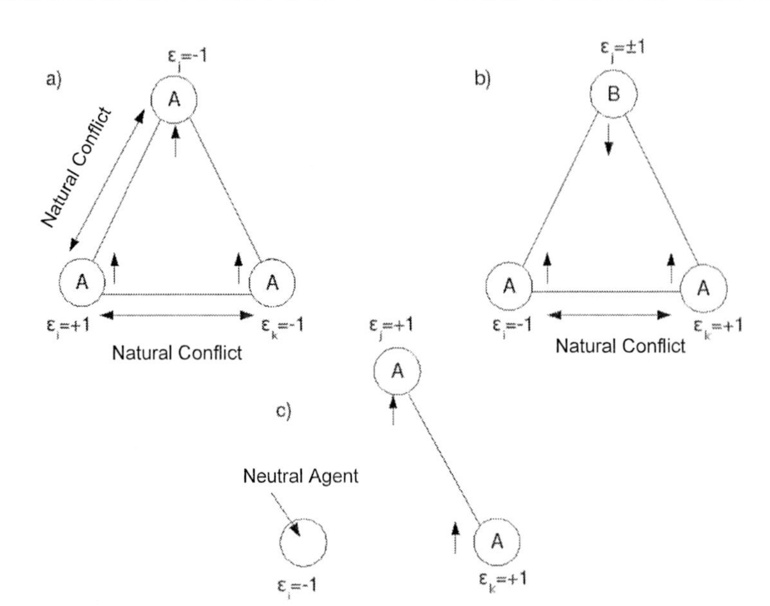

Figure 4. Triangles when there is a natural conflict. The first case is given when all the agents are in the same coalition but their natural propensities are differents (a). The second one is when two actors are in the same coalition and the third is at the opposite one (b). Two agents are in natural conflict with the third one and this one chooses to break his connections staying as a neutral agent (c). The arrows indicate the spin state of the agent, which determines the coalition where he actually is (A or B).

formed by the actors of the system. Thus it is important to make some pertinent modifications to it in order to include the main ingredients of the three-body interactions:

1. When three actors interact between them forming a triangle, a conflict arises if two actors do not have the same natural belongings, given by their corresponding ε_i's. We call this a natural conflict [14]. As a result, the energy must be increased. For example, when the triangle U, M and I is formed, a natural conflict arises due to their different natural belongings (Figure 3.2 a) and b)) . The Equation (11) can be fixed by using a function that is zero when all actors in a triangle have the same natural belongings,

and one in any other case. The corresponding Hamiltonian is,

$$H^{(3)} = \sum_{i,j,k}^{N} \frac{t_{ijk}}{3} \left(\frac{3 - \left|\varepsilon_i + \varepsilon_j + \varepsilon_k\right|}{2} \right) \eta_i \eta_j \eta_k. \tag{13}$$

2. In case of having a natural conflict in a triangle, the increase in energy depends upon the relative configuration of spins. But notice that in Eq. (11), the Hamiltonian is not invariant against the same relative orientation of the spins. For example, the energy of the state $\eta_i = \eta_j = \eta_k = 1$, is not the same as the one obtained from $\eta_i = \eta_j = \eta_k = -1$, although both states have the same relative orientation between them (all parallel). This problem is solved by using the absolute value function,

$$H^{(3)} = \sum_{i,j,k}^{N} \frac{t_{ijk}}{3} \left(\frac{3 - \left|\varepsilon_i + \varepsilon_j + \varepsilon_k\right|}{2} \right) \left|\eta_i \eta_j \eta_k\right|. \tag{14}$$

Since $\eta_i = \pm 1$, $\left|\eta_i \eta_j \eta_k\right| = 1$, it follows that,

$$H^{(3)} = \frac{\left|t_{ijk}\right|}{6} \left(3 - \left|\varepsilon_i + \varepsilon_j + \varepsilon_k\right|\right). \tag{15}$$

3. However, in a natural conflict, one can imagine three configurations: either all actors are in the same coalition (all spins up or down) (Figure 3.2 *a*)), two actors are allied against the third one (Figure 3.2 *b*)), or an actor prefers to "break" the triangle and stays neutral by leaving the network (Figure 3.2 *c*)). We need to assign a penalty in energy for each of these situations, hence it is necessary to insert a function $f(\eta_i, \eta_j, \eta_k)$ in the previous equation (15). In the case of the real Iraq war, the system is more stable when the triangle is broken, instead of trying to build an artificial coalition or a fight. This is the less costly solution, but the penalty is automatically taken into account by $\left|t_{ijk}\right|$ (12), which is zero when the triangle is broken. The next penalty occurs when $\eta_i = \eta_j = \eta_k$; the conflict is solved by an artificial coalition. To assign such penalty with energy W_1, let us first introduce an auxiliary function $f_1(\eta_i, \eta_j, \eta_k)$, with value one when all actors are in the same coalition, and zero in the other case,

$$f_1(\eta_i, \eta_j, \eta_k) = \left(\frac{\left|\eta_i + \eta_j + \eta_k\right| - 1}{2} \right). \tag{16}$$

When two of the actors are in the same coalition and the other is an enemy, we use a function $f_2(\eta_i, \eta_j, \eta_k)$ which is one if two actors are in the same coalition and zero in the other case,

$$f_2(\eta_i, \eta_j, \eta_k) = 1 - f_1(\eta_i, \eta_j, \eta_k) = \left(\frac{3 - |\eta_i + \eta_j + \eta_k|}{2}\right). \tag{17}$$

An energy penalty W_2 is assigned when $f_2(\eta_i, \eta_j, \eta_k) = 1$. Therefore,

$$f(\eta_i, \eta_j, \eta_k) = W_1 f_1(\eta_i, \eta_j, \eta_k) + W_2 f_2(\eta_i, \eta_j, \eta_k)$$

thus the expression for the three-body Hamiltonian is,

$$H^{(3)} = \frac{|t_{ijk}|}{6} \left(3 - |\varepsilon_i + \varepsilon_j + \varepsilon_k|\right) \times \tag{18}$$

$$[W_1 f_1(\eta_i, \eta_j, \eta_k) + W_2 f_2(\eta_i, \eta_j, \eta_k)]. \tag{19}$$

Finally,

$$H^{(3)} = \frac{\delta W}{6} |t_{ijk}| \left(3 - |\varepsilon_i + \varepsilon_j + \varepsilon_k|\right) |\eta_i + \eta_j + \eta_k| + E_r^{(3)}, \tag{20}$$

where $\delta W = (W_1 - W_2)/2$, and $E_r^{(3)}$ is a shift of the energy that only depends on the number of triangles with natural conflicts,

$$E_r^{(3)} = \frac{|t_{ijk}|}{6} \left(3 - |\varepsilon_i + \varepsilon_j + \varepsilon_k|\right) \left(\frac{3W_2 - W_1}{2}\right). \tag{21}$$

The effect of $H^{(3)}$ is to increase the energy of triangles for which a natural conflict is present. If a coalition is artificially set in, it has a penalty W_1, while if the natural conflict is solved by fighting against the common enemy, the penalty is W_2.

3.2. The Application to the Iraq Conflict

In this section, we applied the former mathematical model to the Iraq conflict presented above.

It is important to show how the model behaves according to different values of the parameter α (which determines how strong are the three-body interactions). To make this analysis it is necessary to calculate the energies E_X of every configuration X as a function of this parameter.

To achieve the actual calculation we used the following numerical values for the established parameters in the Hamiltonian (20): $|J'_{ij}| = 0$, $t_{ijk} = C_{ij}C_{ik}C_{jk}$, where $C_F = 10$, $C_E = 3$ and $C = 1$ for all the i, j, k. The value of the weights that penalize each situation are: $W_2 = 1$ and $W_1 = 0$. The fact that we removed W_1 from the scene implies that there is no penalty to insert an artificial coalition. This consideration is valid due to the fact that the term $H^{(3)}$ does not change the energy of the configurations where there are triangles of actors with the same spin state (who belong to the same coalition). The corresponding energy is not changed by α.

Figure 3.3 shows the evolution of the H and $H^{(2)}$ (indicated with arrows) as α is increased. The stars and the diamonds represent the lowest energy state of H and $H^{(2)}$ respectively (considering I and M neutral for the $H^{(2)}$). It is observed in this plot that as the three-body interactions become stronger (α gets larger), there is a critical point α_c where the countries are affected in such a way that the system prefers to disconnect I and M of the network. In other words, when the three-body effects are too strong the countries prefer the neutral state (disconnection).

So far we have presented three models considering the basic ingredients of the coalition forming process. However there is a truly important and delicate aspect within these models [8, 12, 14], which is the proper calculation of the parameters. In the next section we propose some ideas to obtain the natural belongings of the actors in conflict that seek to regroup in the most convenient way to everyone in the system.

4. The Parameters of the Model

The natural belongings are crucial to this model, the configurations which minimize the energy (20) it is strongly correlated with the asignation of the natural belongings ε_i's. These parameters show what is the tendency of the actors to be together before any conflict arises, they are discrete and can take the values $\varepsilon_i = \pm 1$ depending on the coalition which the actor i should belong. The calculation of the natural belongings of a social agent it is a very complex problem, this is because we deal with a great variety of human factors that are difficult to

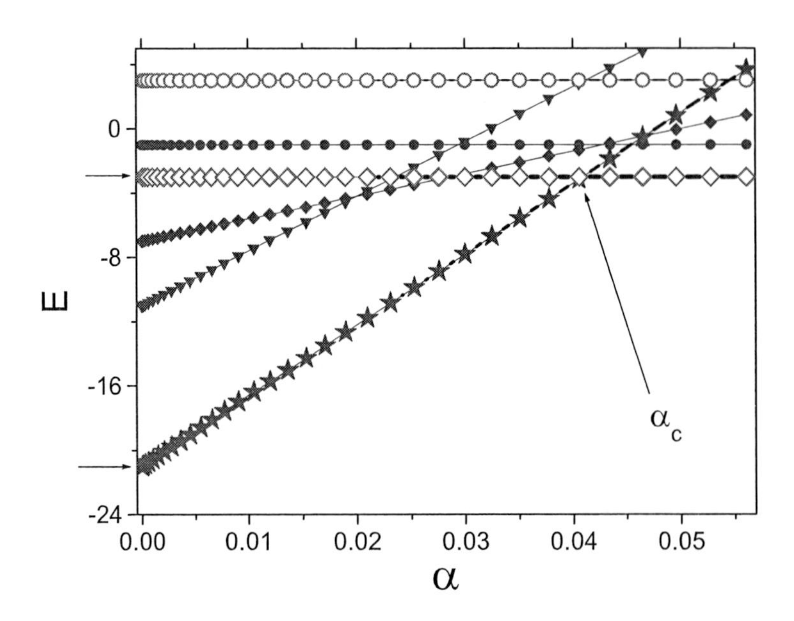

Figure 5. Variation of the ground states of H defined in 20 (stars) and $H^{(2)}$, defined in equation 9 (diamonds), as α is increased. The $H^{(2)}$ Hamiltonian is presented considering I and M as neutral countries (disconnected from the network). Both ground states intersect each other at α_c.

quantifiy such as the economical, social, historical and psychological ones that are very important in the decision making process of countries and societies. The idea here is to find some trends to obtain the parameters, and not to actually present stringent laws which the social agents must follow as those found in the natural sciences.

4.1. The Natural Belongings

As mentioned before, Axelrod and Bennett studied the allignments of the European countries who entered the Second World War (WWII), and Galam and Florian studied the dismembering of the former Yugoslavia. For the first case they calculated the pairwise propensities $p'_{ij}s$ (AB model), and for the second case the natural belongings $\varepsilon_i's$ (G model). We extended their works to develop a three-body model and we applied it to an important geopolitical event such as the recent invasion of Iraq (2003). However their calculation of the parameters could be improved as will be shown below.

Following this previous models, the propensities are important to study the beginning of a conflict because once this has started the previous tendencies of the agents are irrelevant. Collectively the three events discussed by the previous works are similar but at the same time they are different, since the nature and the causes of the conflict were very different. In the first case (WWII) the conflict was studied as an open war between European countries were the axis powers started the war to dominate Europe. In the second situation they studied a different problem; the splitting of a country that led to open warfare. Here the internal divisions and struggles of Yugoslavia were the main factor. Finally, we focused on a slighlty different scenario were the conflict was started by an invasion due to the claim of the production of weapons of mass destruction of a country, together with other goals, like imposing democracy in a dictatorial regime.

To obtain the natural belongings of a country involved in this conflict, we established at least six important aspects that must be considered in the process:

- **Bilateral relations with the USA**: The predominant role in the Iraq invasion was played by the United States of America (USA), and therefore the calculation of the natural belongings naturally involve the bilateral relations of each country with the USA. The high fossil energy consumption of industrialized countries (specially the USA) was a very important factor in this conflict because Iraq ownes vast oil reservoirs.

- **Type of government in each country**: Empirically, it seems that a country prefers to be with another nation with the same type of government. Taking into account the political programs of the governments involved we roughly classified them into ultraleft, left, right and ultra-right wings. This was done considering not only the name of their political, economical and social programs but also by the real situation of the nation and their previous actions.

- **Historical conflicts between the countries**: It is very suitable to think that a country who had previous conflicts with any of the actors of the system, will not share the same natural belongings. Following this argument due to the Gulf War in the early nineties, the USA and Iraq had a different natural belonging.

- **Religion**: If we assume that the people are very religious, this factor is crucial if whether they share or not the same religion in the formation of the alliances.
- **Economical benefits or ties**: We can consider for example the economical dependency between two countries, situation that can be represented in two scenarios: a) one country depends on the other but this is not reciprocal, b) they depend on each other. Following the property that the natural belongings are symmetrical and that the model only works with bilateral relationships (weighted network), we are going to focus on the second scenario only, establishing that the agents who have a great reciprocal economical dependence will have the same natural belonging. This consideration is based under the assumption that if the economical dependency between two countries is very strong, and one of them enters in an international conflict, the other one should belong to the same coalition in order to protect its own interests.
- **Real or imaginary security threats**: A nation will always be prone to belong to the opposite coalition of that one who uses their military weaknesses, in other words if one nation attempt to threaten the national security of another one, they will have different natural belongings in order to protect themselves[2].

We classified the position of each country at the moment of the invasion as being friendly or not friendly to the USA, were the right-wing governments had a strong correlation to align with the USA, as well as the formerly communist countries. In this last group we can observe that these countries were forced to send troops thanks to a three, four or five-body mechanism; this is the case of Poland, who is an historical enemy of Russia and Germany. The situation of

2

- Actually the important aspect to be considered in order to obtain the natural belonging of a nation according to the threat of their national security, is not related with the real threat but actually with the idea that the people of that nation have about it. Following this argument the natural belonging of a nation is determined by the feeling of threat that their people have, no matter if this one is military, economical or religious.

this country have four countries involved (USA, Poland, Germany and Russia), and thanks to their historical conflicts with two of them, the Polish government did the opposite thing that the Germans and the Russians did.

Oil-monarchies (Kuwait, Oman, Katar, United Arab Emirates, Bahrain and Saudi Arabia) are good examples were religion is very important. These countries had a big trend to send troops to the invasion, however this tendency was canceled by the religious factor. This can be contrasted with the first Persian Gulf War when various muslim countries entered to the coalition formed by the USA to attack Iraq. In this case the motivation was the same but the problem was that the Sadam Hussein government attacked Kuwait, another Muslim country. In this example we can observe how the religious beliefs can change the course of a conflict.

4.2. Social Renormalization

We can imagine that societies are built on various hierarchical networks that are interrelated at different scales. Thus, an interesting and very important question in this kind of models lies in the possibility of obtaining the parameters from the internal degrees of freedom of the nations involved. For instance, in the calculation of the natural belongings (which lead to a given position in a conflict), there is a sort of nested social structures -Matryushka like arrangement- that could be modeled in different ways including individuals, families, communities, political parties, and so on.

Notice that this idea was employed in the simplified conflict of the invasion of Iraq analyzed above, were we only considered four agents (U, Q, I, M).

A first attempt in a very simple system was made by Galam [15], where he studied the effect of the majority rule voting on the democratic representation of groups within a hierarchical organization. He rescaled the degrees of freedom to construct hierarchical levels, where in this case the decimated degree of freedom was a real person.

The Renormalization Group Theory (RGT) is a calculational method for establishing important properties in some physical systems (universality, scaling and conformal invariance), as well as to bring a full description of many problems in field theory, statistical physics, dynamical systems theory and many other parts of physics and related subjetcs .

The construction of the RGT lies on the following ideas. The first one is that the scale transformations are important and they may be considered to be a change in the values of the coupling constants. The second is that because of universality (many physical different systems show the same behaviour), different sets of coupling constants may really describe the same physical situation, and an infinite string of them **K** can describe any system. In other words, we can say that a renormalization tranform **R** renormalizes the coupling constants. The final idea is that each phase of the system (especially the critical phase), can be described by a special set of coupling constants, $\mathbf{K}^*$, which are invariant under the renormalization group transform[16]:

$$\mathbf{K}^* = \mathbf{R}(\mathbf{K}^*) \quad criticality\,condition$$

4.3. The Spin Block Example

The easiest example to explain how the RGT works is the spin block one, which is based on the square lattice Ising model. Following the concepts presented in the previous models, we can obtain some ideas in how the social agents at greater scales will interact. An example that shows the importance of this consideration can be seen when the social agents are countries; the interactions between them (exterior policy) are greatly determined by the internal struggles that the nations have, therefore we need to consider these "local" interactions within the agents in order to present a more accurately model.

Lets consider a set of social agents that have divided opinions, one part of them have a position about any subject and the others think the opposite. Now suppose that the opinions are about a conflict with other social agents that are connected to the first set. So, what we would like to know about this situation? What matters is the effective opinion about the conflict that both sets have, and with this information we can know how the sets will interact between them. Lets consider that every social agent can only take one opinion ($s_i = 1$) or its opposite one ($s_i = -1$), that agents only interact with its nearest neighbors (with an energy of J), and that agents are situated in a d dimensional space with N total agents, the energy of the system will be given by

$$H = -J \sum_{\langle i,j \rangle} S_i S_j$$

The $\langle i, j \rangle$ indicates that the addition is made over the nearest neighbors. Now we enclose the social agents in sets by blocks with l^d agents, where $l > 1$

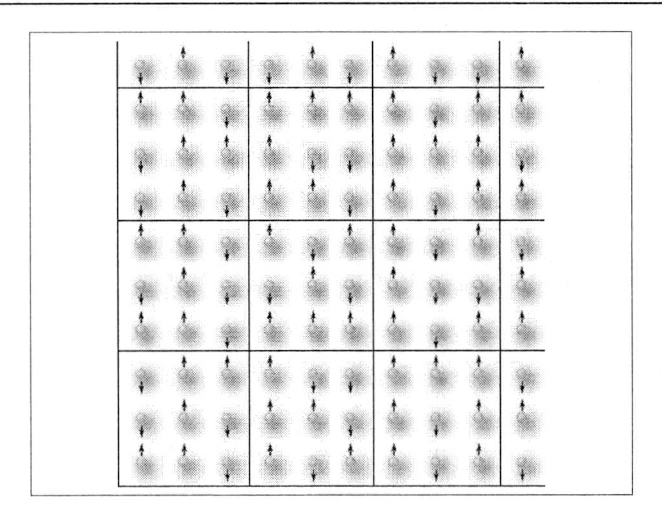

Figure 6. The social agents are in a bidimensional square lattice were each arrow represents an opinion or its opposite. The agents are enclosed in groups of $3^2 = 9$ agents.

(Figure 4.1). Therefore we can calculate the total opinion in each block I with the following expression:

$$S'_I = \sum_{i \varepsilon I} S_i,$$

where the sum is over all the agents in the block.

Now what we have is an effective opinion that will be between $-l^d$ and l^d; but what we actually want is that the block have an opinion with the same nature as that one showed by the original agents. Thus the best thing to do is to rescale this S'_I in the blocks as follows:

$$S'_I = sign(S_I),$$

where $sign(x)$ is the sign of x.

Originally we had agents that interact only with their closest neighbors, now we have blocks of them with the same property but they will interact with an energy of J_l instead. Therefore the new effective Hamiltonian is:

$$H_l = -J_l \sum_{I \neq J}^{N/l^d} S_I S_J$$

In principle J_l can be calculated from the original set of J. Thus, if we know how individuals interact we can obtain in principle how a group of them will interact with the closest groups. This scaling can be done several times in order to obtain the effective interactions between larger groups of social agents. In other words, if we want to know how two countries are related according to their citizens, we need to rescale the parameters more than once considering groups of people, communities, states and so on.

A good example of the importance of such internal interactions can be seen in the role played by Mexico in the Iraq invasion. At that time (2003) the Mexican president had a very good relation with the government of the USA, and therefore Mexico had a great tendency to send troops to Iraq. However, thanks to the people that was against an involvement in the conflict, this country did not participate in the war. Thus, it is important to consider that the final decision to enter into a war depends on the whole collective social structure.

5. Conclusion

In general, a succesful model of coalition formation process can not be made without considering many-body effects. Also, some concepts about the calculation of the main parameters were presented considering mainly economical, social and political factors. In particular we applied these ideas to the coalition forming process in the recent Iraq conflict. We developed a first attempt to consider many-body interactions in the formation of coalitions.

A systematic method to consider the internal degrees of freedom can be the usage of a Renormalization Group formalism as in physics.

We assumed that the propensities between social agents are symmetrical, although asymmetry could be important. When agents are people, it is common to find affection or hate relationships that are asymmetrical.

Although the model certainly captures the principal features of the conflict, the main debate to apply it is still the evaluation and the number of the parameters involved. In reality, the nature of these parameters is fuzzy, that is, they can not be calculated exactly because the error involved in its "measurement" could be large. The behavior of political leaders is a very good example of this because their actions can not be predicted accurately. Some of these effects can also be modeled by introducing a stochastic behavior, analogous to the temperature in physical systems. High temperatures makes the individual behavior less deterministic.

References

[1] Ring P.; Schuck P. The Nuclear Many-Body Problem; Springer: Berlin, 2003; pp 147-640.

[2] Beyer M.; Mattiello S.; Frederico T.; Weber H. *J. Phys Lett B* 2001, 521, 33.

[3] Boyle J. J.; Pindzola M. S. *Many-body atomic physics;* Cambridge University Press: Cambridge, 1998; pp 193-123.

[4] Tuzun R. E.; Noid D. W.; Sumpter B. G. *Journal of Computational Chemistry* 1997, 18, 1513 .

[5] Saksena R. S.; Woodcock L. V.; Maguire J. F. *Mol Phys* 2004, 102, 259.

[6] Saperstein A. M. *Conflict management and peace science* 2004, 21, 287 .

[7] Lambiotte R.; Ausloos M. *Phys Rev E* 2005, 72, 66-117.

[8] Axelrod R., & Bennett D. S. (1993), A Landscape Theory of Aggregation. *British J. Political Sciences,* 23 (2), 211-233.

[9] Axelrod R., Mitchell W., Thomas R. E., Bennett S., & Bruderer E.R. (1995). Coalition Formation in Standard-Setting Alliances. *Management Sci.* **41**, 17-32.

[10] Florian R.; Galam S. *Eur Phys J B* 2000, 16, 189-194.

[11] Binder K.; Young A. P. *Review of Modern Physics* 1986, 58, 801.

[12] Galam S. *Phys A* 1996, 230, 174-188.

[13] Galam S. *Lectures notes in computer science* 2002, 2493.

[14] Naumis Gerardo G.; Samaniego-Steta F.; Del Castillo-Mussot M.; Vazquez G. *J. Phys A* 2007, 379, 226-234.

[15] Galam S. *Phys A* 1999, 274, 132-139.

[16] Kadanoff L. P. *Statistical Physics: Statics, Dynamics & Renormalization;* World Scientific: Singapore, 2000; pp 226-257.

In: Perspectives in Applied Mathematics
Editor: Jordan I. Campbell, pp. 27-41
ISBN: 978-1-61122-796-3

Chapter 2

SUBSTITUTABLE AND PERISHABLE INVENTORY SYSTEM WITH PARTIAL BACKLOGGING

N. Anbazhagan[1,*], C. Elango[2] and P. Subramanian[3]
[1]Department of Mathematics, Alagappa University, Karaikudi, India.
[2]Department of Mathematical Sciences,
Cardamom Planters' Association College, Bodinayakanur, India.
[3]Department of Mathematics,
Thiagarajar College of Engineering, Madurai, India.

Abstract

This article presents a two commodity stochastic perishable inventory system under continuous review. The maximum storage capacity for the i-th item is fixed as S_i $(i = 1,2)$. It is assumed that demand for the i-th commodity is of unit size and independent of the other commodity. Demand time points form a Poisson process with parameter λ_i, $i = 1,2$. The life time of each item of i-th commodity is exponential with parameter γ_i, $i = 1,2$. The reorder level is fixed as s_i for the i-th commodity $(i = 1,2)$ and the ordering

[*] The author is supported by DST - Fast Track Scheme grant for ``Young Scientists", Govt. of India through research project SR/FTP/MS-04/2004.

policy is to place order for $Q_i(=S_i - s_i)$ items of the i-th commodity $(i=1,2)$ when both the inventory levels are less than or equal to their respective reorder levels. The lead time for replenishment is assumed to be exponential with parameter μ. The two commodities are assumed to be substitutable. That is, if the inventory level of one commodity reaches zero, then any demand for this commodity will be satisfied by the other commodity. If substitute is not available, then this demand is backlogged up to a certain level N_i, $(i=1,2)$ for the i-th commodity. Whenever the inventory level reaches N_i, $(i=1,2)$, an order for N_i items is placed and replenished instantaneously. For this model, the limiting probability distribution for the joint inventory levels is computed. Various operational characteristics and expression for long run total expected cost rate are derived.

1. Introduction

The modelling of multi-item inventory system under joint replenishment has been receiving considerable attention for the past three decades. These systems unlike those dealing with single commodity involve more complexities in the reordering procedures. In many practical multi-item inventory systems researchers concentrated on the coordination of replenishment orders for group of items. It is very much applicable to run a successful business.

For continuous review inventory systems, Ballintfy [1964] and Silver[1974] have considered a coordinated reordering policy which is represented by the triplet (S,c,s), where the three parameters S_i, c_i and s_i are specified for each item i with $s_i \le c_i \le S_i$, $(i=1,2)$ under the unit sized Poisson demand and constant lead time. In this policy, if the level of i-th commodity at any time is reaches s_i, an order is placed for $S_i - s_i$ items and at the same time, any other item $j(\neq i)$ with available inventory level at or below its can-order level c_j, an order is placed so as to bring its level back to its maximum capacity S_j. Subsequently many articles have appeared with models involving the above policy and another article of interest is due to Federgruen, Groenevelt and Tijms [1984], which deals with the general case of compound Poisson demands and non-zero lead times. A review of inventory models under joint replenishment is provided by Goyal and Satir[1989].

Kalpakam and Arivarignan [1993] have introduced (s, S) policy with a single reorder level s defined in terms of the total number of items in the stock. This policy avoids separate ordering for each commodity and hence a single processing of orders for both commodities has some advantages in situation where in procurement is made from the same supplies, items are produced on the same machine, or items have to be supplied by the same transport facility.

Krishnamoorthy, Iqbal Basha and Lakshmy [1994] have considered a two commodity continuous review inventory system without lead time. In their model, each demand is for one unit of first commodity or one unit of second commodity or one unit of each commodity 1 and 2, with prefixed probabilities. Krishnamoorthy and Varghese [1994] have considered a two commodity inventory problem with instantaneous replenishment and with Markov shift in demand for the type of commodity namely "commodity-1", "commodity-2" or "both commodity", using the direct Markov renewal theoretical results. And also for the same problem, Sivasamy and Pandiyan [1998] had derived various results by the application of filtering technique.

Anbazhagan and Arivarignan [2000] have considered a two commodity inventory system with Poisson demands and a joint reorder policy which placed fixed ordering quantities for both commodities whenever both inventory levels are less than or equal to their respective reorder levels.

Anbazhagan and Arivarignan [2001] have analyzed models with a joint ordering policy which places orders for both commodities whenever the total net inventory level drops to a prefixed level s.

Yadavalli, Anbazhagan and Arivarignan [2004] have analysed models with individual and joint ordering policy. For the individual reorder policy, the reorder level for i-th commodity is fixed as r_i and whenever the inventory level of i-th commodity falls on r_i an order for P_i $(= S_i - r_i)$ items is placed for that commodity irrespective of the inventory level of the other commodity. A joint reorder policy is used with prefixed reorder levels s and order for Q_x^1 $(= S_1 - x)$ and Q_y^2 $(= S_2 - y)$items is placed for both commodities by cancelling the previous orders, whenever both commodities have their inventory level drops to a reorder level s, $(x + y = s)$.

Anbazhagan [2006] have considered a two commodity substitutable inventory system. If no substitute is available, then this demand is backlogged. The backordering is allowed upto the level N_i, $(i = 1,2)$ for the $i-$th commodity.

Whenever the inventory level reaches N_i, $(i = 1,2)$, an order for N_i items are placed which is replenished instantaneously. Insituations the inventory level raise to the level 0 and then $Q_i, i = 1,2$.

In this paper, we consider an inventory system in which items are perishable in nature. The demand points for each commodity form independent Poisson process and the lead times initiated by joint reorder policy are assumed to be independent and distributed as negative exponential. The two commodities are assumed to be substitutable. That is, if the inventory level of one commodity reaches zero, then any demand for this commodity will be satisfied by an item of the other commodity. If no substitute is available, then this demand is backlogged. The backlog is allowed upto the level N_i, $(i = 1,2)$ for the $i-$th commodity. Whenever the inventory level reaches N_i, $(i = 1,2)$, an order for N_i items are placed which is replenished instantaneously. The limiting probability distribution of the joint inventory level is derived and various measures of system performance in the steady state are also obtained.

2. Model Description

Consider a two commodity stochastic perishable inventory system with the maximum capacity S_i units for i-th commodity $(i = 1,2)$. The demand for i-th commodity is of unit size and the time points of demand occurrences form independent Poisson processes each with parameter λ_i, $(i = 1,2)$. The life time of each item of i-th commodity is exponential with parameter γ_i, $i = 1,2$. The two commodities are assumed to be substitutable. That is, if the inventory level of one commodity reaches zero, then any demand for this commodity will be satisfied by the item of the other commodity. If no substitute is available that is the inventory level is $(0,0)$, then demands are backlogged respectively. The backlog is allowed only up to the level N_i, $(i = 1,2)$ for the i-th commodity. Whenever the inventory level reaches N_i, $(i = 1,2)$, an order for N_i items are placed for instantaneous replenishment. The reorder level for the i-th commodity is fixed at s_i, $(1 \le s_i \le S_i)$ and ordering quantity for i-th commodity is Q_i $(= S_i - s_i > s_i + N_i)$ items, when both inventory levels are less than or equal to their respective reorder levels. The requirement $S_i - s_i > s_i + N_i$, ensures that after a replenishment the inventory levels of both commodities will be always above the respective reorder levels. Otherwise it may not be possible to initiate next reorder which leads to perpetual shortage. That is, if $L_i(t)$ represents inventory level of i-th commodity at time t, then a reorder is made when $L_1(t) \le s_1$ and $L_2(t) \le s_2$. The lead time is assumed to be distributed as negative exponential with parameter $\mu(>0)$.

Notations

0 : zero matrix
$1'_N$: $(1,1,\cdots,1)_{1\times N}$
I_N : an identity matrix of order N
δ_{ij} : Kronecker delta.
$\sum_{k=i}^{j} a^k = \begin{cases} a^i + a^{i+1} + \cdots + a^j, & if\ i<j \\ 0, & otherwise \end{cases}$
$[A]_{ij}$: (i,j) – th element of the matrix A.
$H(x) = \begin{cases} 1 & if\ x \ge 0 \\ 0 & if\ x < 0 \end{cases}$

3. Analysis

From the assumptions made on demand, perishability and replenishment processes, it follows that $\{(L_1(t), L_2(t)), t \ge 0\}$ is a Markov process with state space

$$E = E_1 \cup E_2 \cup E_3,$$

where

$$E_1 = \{(i,j) \mid i = 1,2\cdots,S_1, j = 0,1,\cdots,S_2\},$$

$$E_2 = \{(i,j) \mid i = 0, j = -(N_2-1),-(N_2-2),\cdots,0,1,\cdots,S_2\},$$

and

$$E_3 = \{(i,j) \mid i = -(N_1-1),-(N_1-2),\cdots,-1, j = -(N_2-1),-(N_2-2),\cdots,-1,0\}.$$

The infinitesimal generator $\tilde{A} = ((\, a((i,j),(k,l))\,));\quad (i,j),(k,l) \in E,$ of this process can be conveniently expressed as a block partitioned matrix:

$$\tilde{A} = ((A_{ij})),$$

where

$$A_{ij} = \begin{cases} E_2 & j = i,\ i = -(N_1-1),-(N_1-2),\ldots,-1 \\ E_1 & j = i,\ i = 0 \\ A_i & j = i,\ i = 1,2,\ldots,S_1 \\ D_3 & j = i + N_1 - 1,\ i = -(N_1-1) \\ D_2 & j = i-1,\ i = -(N_1-2),\ldots,-1 \\ D_1 & j = i-1,\ i = 0 \\ B_1 & j = i-1,\ i = 1 \\ B & j = i-1,\ i = 2,3,\ldots,S_1 \\ C_2 & j = i + Q_1, i = -(N_1-1),-(N_1-2),\ldots,-1 \\ C_1 & j = i + Q_1, i = 0, \\ C & j = i + Q_1, i = 1,2,\ldots,s_1, \\ \mathbf{0} & \mathit{Otherwise}. \end{cases}$$

More explicitly $\tilde{A}$ can be written as,

$$
\begin{array}{c}
\\
S_1 \\
S_1-1 \\
\vdots \\
s_1+1 \\
s_1 \\
s_1-1 \\
\vdots \\
1 \\
0 \\
-1 \\
\vdots \\
-(N_1-2) \\
-(N_1-1)
\end{array}
\left(\begin{array}{cccccccccccc}
A_{S_1} & B_{S_1} & & & & & & & & & & \\
 & A_{S_1-1} & B_{S_1-1} & & & & & & & & & \\
 & & \cdots & & & & & & & & & \\
 & & \cdots & & & & & & & & & \\
 & & & A_{s_1+1} & B_{s_1+1} & & & & & & & \\
C & & & & A_{s_1} & B_{s_1} & & & & & & \\
 & \cdots & & & & \cdots & & & & & & \\
 & \cdots & & & & \cdots & & & & & & \\
 & & C & & & & A_1 & B_1 & & & & \\
 & & & C_1 & & & & E_1 & D_1 & & & \\
 & & & & C_2 & & & & E_2 & D_2 & & \\
 & & \cdots & & & & & & & \ddots & & \\
 & & & & & & C_2 & & & & E_2 & D_2 \\
 & & & & & & & C_2 & \cdots D_3 & & & E_2
\end{array}\right)
$$

$$
\text{where } [C_2]_{pq} = \begin{cases} \mu, & q = p + Q_2, \quad p = -(N_2-1), -(N_2-2), \ldots, -1, 0, \\ 0, & otherwise \end{cases}
$$

$$
[C_1]_{pq} = \begin{cases} \mu, & q = p + Q_2, \quad p = -(N_2-1), \ldots, 0, \ldots, s_2 \\ 0, & otherwise \end{cases}
$$

$$
[C]_{pq} = \begin{cases} \mu, & q = p + Q_2, \quad p = 0, 1, \ldots, s_2 \\ 0, & otherwise \end{cases}
$$

$For\ i = 1,2,3$

$$
[D_i]_{pq} = \begin{cases} \lambda_1, & q = p, \qquad p = -(N_2-1), -(N_2-2), \ldots, -1, 0, \\ 0, & otherwise \end{cases}
$$

$For\ i = 1,2,\cdots,S_1$

$$[B_i]_{pq} = \begin{cases} \lambda_1 + i\gamma_1, & q = p, \quad p = 1,2,\ldots,S_2, \\ \lambda_1 + \lambda_2 + i\gamma_1, & q = p, \quad p = 0, \\ 0, & otherwise \end{cases}$$

$$[E_2]_{pq} = \begin{cases} \lambda_2, & q = p-1, \quad p = -(N_2-2),\ldots,-1,0, \\ \lambda_2, & q = p + N_2 - 1, \quad p = -(N_2-1) \\ -(\lambda_1+\lambda_2+\mu), & q = p, \quad p = -(N_2-1),\ldots,-1,0, \\ 0, & otherwise \end{cases}$$

$$[E_1]_{pq} = \begin{cases} \lambda_1+\lambda_2+p\gamma_2, & q = p-1, \quad p = 1,2,\ldots,S_2, \\ \lambda_2, & q = p-1, \quad p = -(N_2-2),\ldots,-1,0, \\ \lambda_2, & q = p+N_2-1, \quad p = -(N_2-1) \\ -(\lambda_1+\lambda_2+\mu), & q = p, \quad p = -(N_2-1),\ldots,0 \\ -(\lambda_1+\lambda_2+\mu+p\gamma_2), & q = p, \quad p = 1,2,\ldots,s_2 \\ -(\lambda_1+\lambda_2+p\gamma_2), & q = p, \quad p = s_2+1,\ldots,S_2 \\ 0, & otherwise \end{cases}$$

For $i = 1,2,\cdots,s_1$

$$[A_i]_{pq} = \begin{cases} \lambda_2 + p\gamma_2, & q = p-1, \quad p = 1,2,\ldots,S_2, \\ -(\lambda_1+\lambda_2+i\gamma_1+p\gamma_2), & q = p, \quad p = s_2+1,\ldots,S_2, \\ -(\lambda_1+\lambda_2+\mu+i\gamma_1+p\gamma_2), & q = p, \quad p = 0,1,\ldots,s_2, \\ 0, & otherwise \end{cases}$$

For $i = s_1+1, s_1+2,\cdots,S_1$

$$[A_i]_{pq} = \begin{cases} \lambda_2 + p\gamma_2, & q = p-1, \quad p = 1,2,\ldots,S_2, \\ -(\lambda_1+\lambda_2+i\gamma_1+p\gamma_2), & q = p, \quad p = 0,1,\ldots,S_2, \\ 0, & otherwise \end{cases}$$

It may be noted that the matrices $A_i, i=1,2,\cdots,S_1$, $B_i, i=2,3,\cdots,S_1$ and C are of size $(S_2+1)\times(S_2+1)$, E_2 and D_2 are of size $N_2\times N_2$, C_1 is of size $(S_2+N_2)\times(S_2+1)$, C_2 is of size $N_2\times(S_2+1)$, D_3 is of size $N_2\times(S_2+N_2)$, D_1 is of size $(S_2+N_2)\times N_2$, B_1 is of size $(S_2+1)\times(S_2+N_2)$, E_1 is of size $(S_2+N_2)\times(S_2+N_2)$.

It can be seen from the structure of $\widetilde{A}$ that the homogeneous Markov process $\{(L_1(t),L_2(t)),t\geq 0\}$ on the state space E is irreducible, non-null persistent, aperiodic states . Hence the limiting distribution

$$\Phi=\left(\phi^{(S_1)},\phi^{(S_1-1)},\cdots,\phi^{(0)},\cdots,\phi^{-(N_1-1)}\right)$$

with

$$\phi^{(m)}=\begin{cases}\left(\phi^{(m,S_2)},\phi^{(m,S_2-1)},\cdots,\phi^{(m,1)},\phi^{(m,0)}\right), & m=1,2,\cdots,S_1\\ \left(\phi^{(m,S_2)},\phi^{(m,S_2-1)},\cdots,\phi^{(m,-(N_2-1))}\right), & m=0\\ \left(\phi^{(m,-(N_2-1))},\phi^{(m,-(N_2-2))},\cdots,\phi^{(m,0)}\right), & m=-(N_1-1),\cdots,-1\end{cases}$$

where $\phi^{(i,j)}$ denotes the steady state probability for the state (i,j) of the inventory level process, exists and is given by

$$\Phi\widetilde{A}=0 \quad and \sum\sum_{(i,j)\in E}\phi^{(i,j)}=1. \tag{1}$$

The first equation of the above yields the following set of equations:

$$\phi^{(i)}E_2+\phi^{(i+1)}D_2=0, \qquad i=-(N_2-1),-(N_2-2),\cdots,-2,$$

$$\phi^{(i)}E_2+\phi^{(i+1)}D_1=0, \qquad i=-1,$$

$$\phi^{(i)}E_1+\phi^{(i+1)}B_1+\phi^{(-(N_1-1))}D_3=0, \qquad i=0,$$

$$\phi^{(i)} A_i + \phi^{(i+1)} B_{i+1} = 0, \qquad i = 1,2,\cdots,Q_1 - N_1,$$

$$\phi^{(i)} A_i + \phi^{(i+1)} B_{i+1} + \phi^{(i-Q_1)} C_2 = 0, \qquad i = Q_1 - N_1 + 1,\cdots,Q_1 - 1,$$

$$\phi^{(i)} A_i + \phi^{(i+1)} B_{i+1} + \phi^{(i-Q_1)} C_1 = 0, \qquad i = Q_1$$

$$\phi^{(i)} A_i + \phi^{(i+1)} B_{i+1} + \phi^{(i-Q_1)} C = 0, \qquad i = Q_1 + 1,\cdots,S_1 - 1$$

and $$\phi^{(S_1)} A_{S_1} + \phi^{(s_1)} C = 0.$$

We obtain the limiting probability values and express them in the form

$$\phi^i = \phi^{(Q_1)} \theta_i, \quad i = -(N_1 - 1),\cdots,0,1,\cdots,S_1$$

where θ_i values can be obtained recursively from

$$\theta_i = \begin{cases} -\theta_{i+1} D_2 E_2^{-1} & i = -(N_1 - 1),\cdots,-3,-2 \\ -\theta_{i+1} D_1 E_2^{-1} & i = -1 \\ -(\theta_{i+1} B_1 + \theta_{-N_1+1} D_3) E_1^{-1} & i = 0 \\ -\theta_{i+1} B_{i+1} A_i^{-1} & i = 1,2,\cdots,Q_1 - N_1 \\ -(\theta_{i+1} B_{i+1} + \theta_{i-Q_1} C_2) A^{-1} & i = Q_1 - N_1 + 1,\cdots,Q_1 - 1 \\ I & i = Q_1 \\ -(\theta_{i+1} B_{i+1} + \theta_{i-Q_1} C) A_i^{-1} & i = Q_1 + 1,\cdots,S_1 - 1 \\ -\theta_{i-Q_1} C A_i^{-1} & i = S_1 \end{cases}$$

The value of $\phi^{(Q_1)}$ can be obtained from the relation $\sum\sum_{(i,j)\in E} \phi^{(i,j)} = 1$,

as
$$\phi^{(Q_1)} = \left(I + \sum_{\substack{i=-(N_1-1) \\ i \neq Q_1}}^{S_1} \theta_i \right)^{-1}.$$

4. System Performance Measures

In this section we derive some measures of system performance in the steady state under consideration.

Expected Reorder Rate

Let R denote the mean joint reorder rate for both the commodity in the steady state which is given by,

$$R = \sum_{j=0}^{S_2} (\lambda_1 + \delta_{0j} \lambda_2) \phi^{(s_1+1,j)} + \sum_{i=0}^{S_1} (\delta_{i0} \lambda_1 + \lambda_2) \phi^{(i,s_2+1)}.$$

Let R_1 denote the mean individual reorder rate for first commodity in the steady state which is given by,

$$R_1 = \sum_{j=-(N_2-1)}^{0} \lambda_1 \phi^{(-(N_1-1,j))}.$$

Let R_2 denote the mean individual reorder rate for second commodity in the steady state which is given by,

$$R_2 = \sum_{i=-(N_1-1)}^{0} \lambda_2 \phi^{(i,-(N_2-1))}.$$

Expected Perishable Rate

Let PR_1 denote the mean perishable rate of the first commodity in the steady state then we have

$$PR_1 = \sum_{i=1}^{S_1} \sum_{j=0}^{S_2} i\gamma_1 \phi^{(i,j)}.$$

Let PR_2 denote the mean perishable rate of the second commodity in the steady state then we have

$$PR_2 = \sum_{i=0}^{S_1} \sum_{j=1}^{S_2} j\gamma_2 \phi^{(i,j)}.$$

Mean Backlogging

Let B_1 denote the mean backlogging level of the first commodity in the steady state then we have

$$B_1 = \sum_{i=-(N_1-1)}^{-1} \sum_{j=-(N_2-1)}^{0} |i| \phi^{(i,j)}.$$

Let B_2 denote the mean backlogging level of the second commodity in the steady state then we have

$$B_2 = \sum_{j=-(N_2-1)}^{-1} \sum_{i=-(N_1-1)}^{0} |j| \phi^{(i,j)}.$$

Expected Inventory Level

Let I_1 denote the expected inventory level of the first commodity in the steady state which is given by

$$I_1 = \sum_{i=1}^{S_1} i \left(\sum_{j=0}^{S_2} \phi^{(i,j)} \right).$$

Let I_2 denote the expected inventory level of the second commodity in the steady state. Then we have

$$I_2 = \sum_{j=1}^{S_2} j \left(\sum_{i=0}^{S_1} \phi^{(i,j)} \right).$$

Total Expected Cost Rate

To compute the total expected cost rate per unit time, we consider the following costs

h_1	the inventory carrying cost per item per unit time of first commodity.
h_2	the inventory carrying cost per item per unit time of second commodity.
c_s	setup cost per order.
c_{es1}	setup cost per emergency order for first commodity.
c_{es2}	setup cost per emergency order for second commodity.
c_{bi}	backlog cost per unit per unit time of i-th commodity (i=1,2).
c_{pi}	perishable cost per unit item per unit time of i-th commodity (i=1,2).

The long run total expected cost rate is given by

$$TC(S_1, s_1, N_1, S_2, s_2, N_2)$$
$$= h_1 I_1 + h_2 I_2 + c_s R + c_{es1} R_1 + c_{es2} R_2 + c_{b1} B_1 + c_{b2} B_2 + c_{p1} PR_1 + c_{p2} PR_2.$$

5. Conclusion

In this paper we have described a two commodity perishable and substitutable inventory system with partial backlogging. This model is most suitable to two different items which are substitutable like coffee and tea. This work may be extended to the case of renewal demand and stochastic lead time. The major contribution of this model is maintenance strategy for perishable and substitutable commodities in a stochastic demand environment.

References

[1] N. Anbazhagan and G. Arivarignan, Two-Commodity Continuous Review Inventory system with Coordinated Reorder Policy *International Journal of Information and Management Sciences*. **11**(3), 19-30, (2000).

[2] N. Anbazhagan and G. Arivarignan, Analysis of Two-Commodity Markovian Inventory system with lead time. *Journal of Applied Mathematics & Computing,* **8**(2), 519 - 530, 2001.

[3] N. Anbazhagan and G. Arivarignan, Two-Commodity Inventory system with Individual and Joint ordering Policies. *International Journal of Management and Systems*, **19**(2), 129-144, May-August, 2003.

[4] N. Anbazhagan, Two Commodity Substitutable Inventory System with Partial Backlogging, *Proceedings of International Conference on Modeling and Simulation 2006 (MS 2006*), April 03-05, 2006, Department of Mechanical Engineering, University of Malaya, Kulalampur, Malaysia.

[5] J.L.Ballintify, On a basic class of inventory problems. *Management Science.* **10**, 287-297 (1964).

[6] A.Federgruen, H.Groenvelt and H.C.Tijms, Coordinated replenishment in a multi-item inventory system with compound Poisson demands. *Management Science.* **30**, 344-357 (1984).

[7] S.K.Goyal and T.Satir, Joint replenishment inventory control: Deterministic and stochastic models. *European Journal of Operations Research* **38**, 2-13 (1989).

[8] S. Kalpakam and G. Arivarignan, A coordinated multicommodity (s,S) inventory system. *Mathl. Comput. Modelling.* **18**, 69-73 (1993).

[9] A.Krishnamoorthy and N. Anbazhagan, Perishable Inventory system at Service Facilities with N-Policy. *Stochastic Analysis and Applications*, **26**, 120-135, (2008).

[10] A.Krishnamoorthy, R.Iqbal Basha and B.Lakshmy, Analysis of two commodity problem. *International Journal of Information and Management Sciences.* **5**(1), 55-72, (1994).

[11] Krishnamoorthy and T. V. Varghese, A two commodity inventory problem. *Information and Management Sciences*, **5**(3), 127-138, (1994).

[12] E.A.Silver, A control system of coordinated inventory replenishment. *International Journal of Production Research*. **12**, 647-671 (1974).

[13] Sivakumar, N. Anbazhagan and G. Arivarignan, Two-Commodity Continuous Review Perishable Inventory System. *The International Journal of Information and Management Sciences*, **17**(3), 47-64, September 2006.

[14] R. Sivasamy, P. Pandiyan, A two commodity Inventory Model Under (s_k, S_k) Policy, *International Journal of Information and Management Sciences* **9**(3), 19-34, (1998).

[15] V.S.S. Yadavalli, N. Anbazhagan and G. Arivarignan, A Two-Commodity Continuous Review Inventory System with Lost Sales, *Stochastic Analysis and Applications,* **22**, 479 - 497, 2004.

In: Perspectives in Applied Mathematics ISBN 978-1-61122-796-3
Editor: Jordan I. Campbell, pp. 43-50

Chapter 3

THREE-MEMBER COMMITTEE LOOKING FOR A SPECIALIST WITH TWO HIGH ABILITIES

Minoru Sakaguchi
Osaka University, Osaka, Japan

1. Statement and Formulation of the Problem

A 3-player(=member) committee has players I, II and III. The committee wants to employ one specialist among n applicants. It interviews applicants sequentially one-by-one. Facing each applicant player I(II, III) evaluates the management ability at X_1 (Y_1, Z_1) and computer ability at X_2 (Y_2, Z_2). Evaluation by the players are made independently and each player chooses, based on his evaluation, either one of R and A. The committee's choice is made by simple majority. If the committee rejects the first $n-1$ applicants, then it should accept the n-th applicant. Denote

$$\xi = x_1 \wedge x_2 \,, \; \eta = y_1 \wedge y_2 \,, \; \zeta = z_1 \wedge z_2 \,. \tag{1.1}$$

If the committee accepts an applicant with talents evaluated at $\mathbf{x}, \mathbf{y}, \mathbf{z}$ by I, II, III, resp., then the game stops and each player is paid ξ, η, ζ to I, II, III, resp.. If the committee rejects an applicant, then the next applicant is interviewed and the game continues. Each player of the committee aims to maximize the expected payoff he can get.

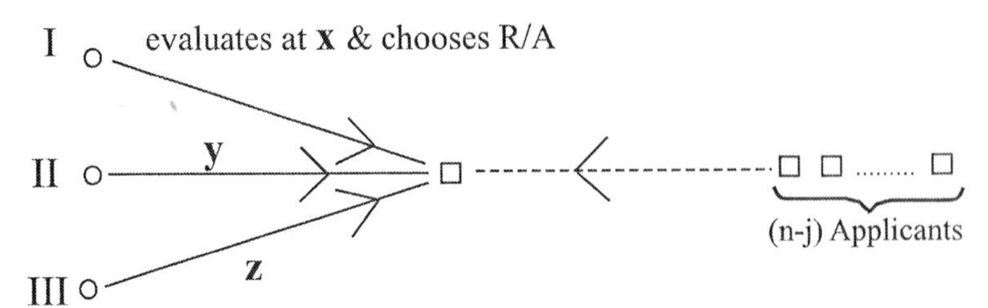

Figure 1. State $(j, \mathbf{x}, \mathbf{y}, \mathbf{z})$.

The two different kinds of talents (management and computer abilities) for each applicant, are bivariate r.v.s, *i.i.d.* with pdf

$$h(x_1,x_2) = 1 + \gamma(1-2x_1)(1-2x_2), \ \forall(x_1,x_2) \in [0,1]^2, \ |\gamma| \leq 1 \qquad (1.2)$$

for player I. For II and III, pdfs are $h(y_1,y_2)$ and $h(z_1,z_2)$ respectively, with the same γ. If X_1 (X_2) for I is the evaluation of ability of management (foreign language), then γ will be $0 \leq \gamma \leq 1$. If X_2 is the evaluation of the computer ability, then γ may be $-1 \leq \gamma \leq 0$.

The bivariate pdf (1.2) is one of the simplest pdf that has the identical uniform marginal and correlated component variables. The correlation coefficient is equal to $\gamma/3$.

Denote the state $(j, \mathbf{x}, \mathbf{y}, \mathbf{z})$ where $\mathbf{x} = (x_1, x_2)$, etc., to mean that 1) the first $j-1$ applicants were rejected by the committee, 2)the j-th applicant is currently evaluated at $\mathbf{x}, \mathbf{y}, \mathbf{z}$, by I, II, III resp. and 3) $n-j$ applicants remain uninterviewed if the j-th is rejected by the committee. The state is illustrated by Figure 1.

We define u_j = Expected payoff, player I can get, if I is in state $(j, \mathbf{x}, \mathbf{y}, \mathbf{z})$ and all players play optimally hereafter.

Define v_j, for II, and w_j, for III, similarly. Moreover we introduce a number

$$c = E_{\mathbf{x}}(\xi) = 2\int_0^1 dx_1 \int_0^{x_1} x_2 h(x_1,x_2)dx_2 = \frac{1}{3} + \frac{1}{30}\gamma$$

where is in $[3/10, 11/30]$ for $\forall \gamma \in [-1,1]$.

The Optimality Equation of our 3-player 2-choice n-stage game is

$$(u_j, v_j, w_j) = E_{\mathbf{x},\mathbf{y},\mathbf{z}}[\text{Optimal payoffs facing } \mathbf{M}_j(x,y,z)], \qquad (1.3)$$

$$(j \in [1,n],\, u_n = v_n = w_n == E_{\mathbf{x}}(\xi) = c)\,,$$

where the payoffmatrix is represented by

$$\mathbf{M}_j(\mathbf{x},\mathbf{y},\mathbf{z}) = \begin{cases} \text{R by I} & \mathbf{M}_{j,R}(\mathbf{x},\mathbf{y},\mathbf{z}) \\ \text{A by I} & \mathbf{M}_{j,A}(\mathbf{x},\mathbf{y},\mathbf{z}) \end{cases} \quad (1.4)$$

$\mathbf{M}_{j,R}(\mathbf{x},\mathbf{y},\mathbf{z}) =$ (1.5)

	R by III	A by III
R by II	$u,\ v,\ w$	$u,\ v,\ w$
A by II	$u,\ v,\ w$	$\xi,\ \eta,\ \zeta$

$\mathbf{M}_{j,A}(\mathbf{x},\mathbf{y},\mathbf{z}) =$ (1.6)

	$u,\ v,\ w$	$\xi,\ \eta,\ \zeta$
	$\xi,\ \eta,\ \zeta$	$\xi,\ \eta,\ \zeta$

because of the simple majority rule.

(In each cell, the subscript $j+1$ of $u_{j+i}, v_{j+i}, w_{j+i}$ is omitted. We use this convention hereafter too, when needed.)

Related problems are investigated in [2, 4, 5] . [1] studies a two player game. [3, 4, 5] are concerned with 3 or more player games.

2. Solution of the Problem

Lemma 1. *For I in state $(j,\mathbf{x},\mathbf{y},\mathbf{z})$, R (A) dominates A (R), if $u_{j+i} > (<)\xi$. By symmetry, for II (III), u_{j+i} and ξ are replaced by v_{j+i} and η (w_{j+i} and ζ).*

Proof. Eq.(1.5) minus Eq.(1.6) is

$\mathbf{M}_R - \mathbf{M}_A =$

	R by III	A by III
R by II	$0,\ 0,\ 0$	$u-\xi,\ v-\eta,\ w-\zeta$
A by II	$u-\xi,\ v-\eta,\ w-\zeta$	$0,\ 0,\ 0$

and we look at the signs of 4 components for I. Hence the lemma follows.

Lemma 2.

$$f(u) = E_{\mathbf{x}}I(\xi > u) = (\overline{u})^2(1+\gamma u^2)\,,\ \forall u \in [0,1]\,. \quad (2.1)$$

This function is decreasing with values $f(0) = 1$, $f(\frac{1}{2}) = \frac{1}{4} + \frac{1}{16}\gamma$, and $f(1) = 0$. Moreover $f(u)$ is convex if $0 < \gamma \leq 1$.

$$g(u) = E_{\mathbf{x}}[\xi I(\xi > u)] = c - u^2 + \frac{2}{3}u^3 + \gamma u^3(\frac{2}{3} - \frac{3}{2}u + \frac{4}{5}u^2) \tag{2.2}$$

is decreasing with values $g(0) = c$, $g(\frac{1}{2}) = \frac{1}{6} + \frac{23}{480}\gamma$, *and* $g(1) = 0$.

Proof.

$$\begin{aligned} f(u) &= \int_u^1 \int_u^1 \{1 + \gamma(1-2x_1)(1-2x_2)\} dx_1 d_2 \\ &= (\bar{u})^2 + \gamma[\int_0^1 (1-2x_1)dx_1]^2 = (\bar{u})^2[1 + \gamma(-u\bar{u})^2] \end{aligned}$$

i.e., Eq.(2.1). Moreover we obtain

$$f''(u) = 2\gamma[\gamma^{-1} + (1 - 6u\bar{u})] > 0 \ \forall u \in [0,1], \text{ if } 0 < \gamma \leq 1.$$

On the other hand

$$\begin{aligned} g(u) = &\int_u^1 dx_1 \int_u^{x_1} x_2 h(x_1,x_2)dx_2 + \int_u^1 dx_2 \int_u^{x_2} x_1 h(x_1,x_2)dx_1 = \\ &2\int_u^1 dx_1 \int_u^{x_1} x_2\{1 + \gamma(1-2x_1)(1-2x_2)\}dx_2 \end{aligned}$$

After a bit of calculations, we have

$$2\int_u^1 (1-2x_1)dx_1 \int_u^{x_1} x_2(1-2x_2)dx_2 = \frac{1}{30} + \frac{2}{3}u^3 - \frac{3}{2}u^4 + \frac{4}{5}u^5$$

and so Eq.(2.2) follows.

Both of $f(u)$ and $g(u)$ are decreasing, because of their definitions.

It is evident that

$$1 > f(u) > g(u) > 0, \ \forall u \in (0,1) \tag{2.3}$$

by the definitions of $f(u)$ and $g(u)$.

Theorem 1. *Optimal expected payoff to I satisfies the recursion*

$$u_j = Q(u_{j+1}),\ \forall j \in [1, n-1],\ u_n = c, \tag{2.4}$$

where

$$Q(u) = u[1 - 3(f(u))^2 + 2(f(u))^3] + (f(u))^2(c - 2g(u)) + 2f(u)g(u). \tag{2.5}$$

In the r.h.s., $f(u)$ and $g(u)$ are given by (2.1) and (2.2) resp., in Lemma 2. $Q(u)$ is a continuous function with values $Q(0) = c$ and $Q(1) = 1$.

Proof. From Lemmas 1 and 2 and Eqs (1.1)-(1.6), the optimal payoff to player I is the sum of eight terms

$$\begin{gathered} u[I(\xi < u, \eta < v, \zeta < w) + I(\xi < u, \eta < v, \zeta > w) + I(\xi < u, \eta > v, \zeta < w) + \\ + I(\xi > u, \eta < v, \zeta < w) + I(\xi < u, \eta > v, \zeta > w) + I(\xi > u, \eta < v, \zeta > w) + \\ + I(\xi > u, \eta > v, \zeta < w) + I(\xi > u, \eta > v, \zeta > w)]. \end{gathered}$$

Remember that the u, v, w, here, are $u_{j+i}, v_{j+i}, w_{j+i}$, resp.

Committee's decision is R (A) in the first (second) 4 events.

Taking $E_{\mathbf{x},\mathbf{y},\mathbf{z}}$ of the r.v.s, we get

$$\begin{gathered} u\{\overline{f(u)}\,\overline{f(v)}\,\overline{f(w)} + \overline{f(u)}\,\overline{f(v)}\,f(w) + \overline{f(u)}\,f(v)\,\overline{f(w)} + f(u)\,\overline{f(v)}\,\overline{f(w)}\} + \\ + \{(c - g(u))f(v)f(w) + g(u)\overline{f(v)}f(w) + g(u)f(v)\overline{f(w)} + g(u)f(v)f(w)\}. \end{gathered}$$

From symmetry among the three players, we can take $u = v = w$. Then the above expression becomes

$$\begin{gathered} u\{\overline{f(u)}^3 + 3f(u)\overline{f(u)}^2\} + \{c(f(u))^2 + 2g(u)f(u)\overline{f(u)}\} = \\ u\overline{f}(u)^2\{\overline{f(u)} + 3f(u)\} + c(f(u))^2 + 2g(u)f(u)\overline{f(u)} = \\ u\{1 - 3(f(u))^2 + 2(f(u))^3\} + (f(u))^2(c - 2g(u)) + 2f(u)g(u). \end{gathered}$$

The last expression is $Q(u)$ given by (2.5).

Theorem 2.

$$u_1 > u_2 > \cdots u_n = c \tag{2.6}$$

Proof. We have by Theorem 1,

$$\begin{aligned} u_j - u_{j+1} &= Q(u_{j+1} - u_{j+1}) = \\ &= [u\{-3(f(u))^2 + 2(f(u))^3\} + (f(u))^2(c - 2g(u)) + 2f(u)g(u)]_{u=u_{j+1}} = \\ &= f(u_{j+1})[u\{-3f(u) + 2(f(u))^2\} + (f(u))(c - 2g(u)) + 2g(u)]_{u=u_{j+1}}. \end{aligned}$$

We want to prove that the inside of $[\cdots]$, i.e.,

$$m(u) = 2g(u) + f(u)(-3u + -2g(u)) + 2u(f(u))^2 \tag{2.7}$$

is positive for $\forall u \in (0,1)$. We find that its proof is un-expectedly intractable. Clearly $m(0) =$ and $m(1) = 0$.

Suppose that $m(u_0) = 0$ for some $u_0 \in (0,1)$. Then, from (2.4) and (2.7), it must hold

$$\begin{aligned} (\bar{u_0})^2(1+\gamma u_0^2) &= f(u_0) = \tfrac{3}{4} + (4u_0)^{-1}[2g - c - \sqrt{(3u_0 - c + 2g)^2 - 16u_0 g}] = \\ &= \tfrac{3}{4} + (4u_0)^{-1}[2g - c - \sqrt{(2g-c)^2 + (c - 3u_0)^2 - (4u_0 g + c^2)}] \end{aligned} \tag{2.8}$$

The inside of the square root becomes negative for $u_0 = (1/3)$, since

$$(2g-c)^2 - (\frac{4}{3}cg + c^2) = 4g(g - \frac{4}{3}c) < 0\,.$$

So, Eq.(2.8) doesn't hold true.

Hence $m(u) \neq 0$, $\forall u \in (0,1)$, and since $m(0) = c$, and $m(1) = 0$, we find that $m(u) > 0$, $\forall u \in (0,1)$.

3. The Case $\gamma = 0$

Consider the special case $\gamma = 0$. Then from (2.1), (2.2) and (2.4) we obtain

$$c_0 = \frac{1}{3},\ f_0(u) = \bar{u}^2,\ g_0(u) = \frac{1}{3} - u^2 + \frac{2}{3}u^3,$$

and

$$\begin{aligned} Q_0(u) &= u[1 - 3(f_0(u))^2 + 2(f_0(u))^3] + (f_0(u))^2(\tfrac{1}{3} - 2g_0(u)) + 2f_0(u)g_0(u) = \\ &= u + \bar{u}^2(\tfrac{2}{3} - 2u^2 + \tfrac{4}{3}u^3) + \bar{u}^4(-\tfrac{1}{3} - 3u + 2u^2 - \tfrac{4}{3}u^3) + 2u\bar{u}^6\,. \end{aligned} \tag{3.1}$$

Therefore

$$\begin{aligned} u_{n-1} &= Q_0(1/3) = \tfrac{1}{3} + \tfrac{4}{9}\tfrac{40}{81} + \text{ two more terms} \\ &\approx \tfrac{1}{3} + 0.21948 - 0.22923 + 0.05853 \approx 0.38211 \end{aligned} \tag{3.2}$$

and

$$u_{n-2} = Q_0(0.38211) \approx 0.3821 + 0.17145 - 0.18397 + 0.04253 \approx 0.4121\,.$$

We have a conjecture that the sequence (2.6) is concavely decreasing.

An example.

There are 3-player committee and 3-applicants. The common optimal strategy for each player is

"Choose A (R), if $x_1 \wedge x_2 > (<) u_1 = 0.4121$ in the first stage" [For player II (III) $x_1 \wedge x_2$ is replaced by $y_1 \wedge y_2 (z_1 \wedge z_2)$.]

If the committee rejects the 1st applicant, then it interviews the 2nd applicant, and

"Choose A (R), if his $x_1 \wedge x_2 > (<) u_2 = 0.3821$ in the second stage"

If the committee rejects the 2nd applicant, then the committee should accept the last applicant. Each player's expected payoff is $u_3 = 1/3$.

4. Final Remark

The present author cannot use computer by some inevitable private reasons. If we can use computer, it would be interesting to make the table of

	u_n	u_{n-1}	u_{n-2}	$\cdots$
$\gamma = -1$	3/10	0.3420	0.3685	$\cdots$
0	1/3	0.3821	0.4121	$\cdots$
1	11/30	0.4284	0.4620	$\cdots$

(The numbers above are obtained, from (2.1)-(2.5), by using a small calculator.)

References

[1] F. Ben Abdelaziz, S. Krichen, An interactive method for the optimal selection problem with two decision makers, *Europ. J. Oper. Res.*, (to appear).

[2] T. Ferguson, Selection by committee, *Annals of the International Society of Dynamic Games* 7 (2005), Advances in Dynamic Games Application to Economics, Finance, Optimization and Stochastic Control, 203–209.

[3] V. V. Mazalov, M. V. Banin, N person best choice games with voting, *Game Th. Appl.* **IX** (2003), 45–53.

[4] V. V. Mazalov, M. Sakaguchi, A. A. Falco, *Selection by committee in the best-choice problem with rank criterion*, to appear.

[5] M. Sakaguchi, *Three-member committee where odd-man's judgement is paid regard*, to appear.

In: Perspectives in Applied Mathematics
Editor: Jordan I. Campbell, pp. 51-73
ISBN 978-1-61122-796-3

Chapter 4

Contractual Stability and Competitive Equilibrium in a Pure Exchange Economy

Valery A. Vasil'ev
Sobolev Institute of Mathematics
Russian Academy of Sciences, Siberian Branch, Russia

1. Introduction

This paper contains a game-theoretical analysis of the so-called weak totally contractual allocations, similar to that introduced by V.L.Makarov [4] in order to describe stable outcomes of some quite natural recontracting processes in pure exchange economies. Detailed presentation of Makarov's original settings can be found in [5]. Below, we analyze a slightly strengthened version of contractual blocking, introduced in this paper. Namely, in what follows, no additional restrictions to the stopping rule of the breaking procedure is posed besides the feasibility of the final contractual system (hence, no minimality condition, applied in [4-6], and [8,9]). At the same time, like in [4-6], any such a final contractual system supposed to be an improvement for each member of blocking coalition (not just at least one of the minimal final system, like it appears to be in [8,9]).

So, being close to the original definition of contractual blocking, the version exploiting in the paper seems to be quite far from that introduced in [9]. In fact, even for the recontracting process, investigated in the latter paper, the multiplicity of outcomes of contractual breaking procedure is quite typical. Remind [9], that it may often be the case that after by some coalition chosen contracts are broken, the rest (including newly concluded by this coalition) do not constitute a feasible system. It is inherent in the model that in such a situation the breaking process proceeds spontaneously, and stops after feasibility of the contract system is recovered. The only requirement we deal with on this step is minimality: the spontaneous process breaks (nullifies) collection of contracts as small, as possible, provided that not nullified ones constitute a feasible contract system. It is clear that by omitting the minimality condition one enlarges the number of outcomes of the breaking procedure considerably. Therefore, to convince the contractual blocking exploiting in the paper is much more stronger than that applied in [9], we only stress once more that the former blocking requires blocking coalition to improve upon the initial contract system with any possible breaking outcome (not just with one of them, like in [9]).

Surprisingly, for several rather wide classes of markets, the cores corresponding to the above mentioned types of blocking, are the same. To clarify this phenomenon, we prove one of the main results of the paper, stating that under rather mild assumptions the weak totally contractual core (consisting, by definition, of the allocations that are stable w.r.t the strengthened blocking) is equal to the set of Walrasian equilibrium allocations. To demonstrate the main assumptions in the core equivalence theorem, mentioned above, are relevant, two examples of pure exchange economies having unblocked allocations with no supporting equilibrium prices are given. The most interesting seems to be the last example with no equilibrium allocations and nonempty weak totally contractual core, exhibiting that a weak totally contractual allocation may be chosen as a compromise solution in case the classical market mechanism doesn't work.

2. Weak Totally Contractual Core

Below, we consider a slightly generalized pure exchange model

$$\mathcal{E} = \langle N, \{X_i, w^i, \alpha_i\}_N, \sigma \rangle, \tag{1}$$

where $N := \{1, \ldots, n\}$ is a set of consumers, and $X_i \subseteq \mathbf{R}^l$, $w^i \in \mathbf{R}^l$, $\alpha_i \subseteq X_i \times X_i$ are their consumption sets, initial endowments, and individual preference relations, respectively. As to σ, which is a nonempty subset of 2^N, called a coalitional structure of $\mathcal{E}$, it defines a collection of admissible coalitions $S \subseteq N$, which may join efforts of their participants in order to improve (block) any current allocation of the total initial endowment $\sum_N w^i$.

Remind also [4], that in our notations inclusion $(x,y) \in \alpha_i$ means that y is more preferable than x (w.r.t. the preference relation α_i). For any $x \in X_i$, we denote by $\alpha_i(x)$ a collection of all the bundles $y \in X_i$ that are more preferable than x

$$\alpha_i(x) := \{y \in X_i \mid (x,y) \in \alpha_i\}.$$

As usual, we apply the following shortenings: $x\alpha_i y \Leftrightarrow (x,y) \in \alpha_i$, and

$$\mathcal{P}_i(x) := \{y \in X_i \mid y \in \alpha_i(x), x \notin \alpha_i(y)\}$$

(recall, that in the standard interpretation, inclusion $y \in \mathcal{P}_i(x)$ means that y is strictly more preferable than x).

We present first the main notion of the weak contractual domination (blocking) in $\mathcal{E}$. In order to do that, we modify first some relevant definitions from [6], aiming to clarify the main features of the elementary interchange structure we deal with. For each $S \in \sigma$ we fix some subset $M_S \subseteq \mathbf{R}^l$ such that $0 \in M_S$, and put $M_\sigma := \{M_S\}_\sigma$.

Definition 1. *A contract (of type M_S) of coalition $S \in \sigma$ is a collection of vectors* $\mathrm{v} = \{\mathrm{v}^{ij}\}_{i,j \in S}$ *satisfying conditions:*

(a) $\mathrm{v}^{ii} = 0$ *for any* $i \in N$,

(b) $\mathrm{v}^{ij} \in M_S$ *for any* $i, j \in S$,

(c) $\mathrm{v}^{ij} = -\mathrm{v}^{ji}$ *for any* $i, j \in S$.

Coalition S entering into a contract v will be denoted by $S(\mathrm{v})$, as well.

Components v^{ij}, appearing in definition of the contract $\mathrm{v} = \{\mathrm{v}^{ij}\}_{i,j \in S}$, indicate amount of the corresponding commodities used in the bilateral exchange between the participants $i, j \in S(\mathrm{v})$. Here nonnegative component $\mathrm{v}_k^{ij} \geq 0$ of the vector v^{ij} denotes the amount of commodity k that agent j is obliged to deliver to agent i, and absolute value of negative component $\mathrm{v}_r^{ij} < 0$ measures the amount of commodity r agent i has to deliver to agent j.

As to the subsets M_S, they define admissible types of elementary exchanges within the contract $\mathrm{v} = \{\mathrm{v}^{ij}\}_{i,j \in S}$, e.g., in case $M_S := \{x \in \mathbf{R}^l \mid p^S \cdot x = 0\}$ the only constraint concerning v^{ij}, $i,j \in S$ is that the bilateral exchanges v^{ij} should be equivalent w.r.t. the (fixed) prices p^S.

We call $\mathrm{v} = \{\mathrm{v}^{ij}\}_{i,j \in S}$ *a proper contract*, if either v is a trivial contract (i.e., $\mathrm{v}^{ij} = 0$ for all $i,j \in S(\mathrm{v})$), or for any $i \in S(\mathrm{v})$ there exist $j,k \in S(\mathrm{v})$ such that v^{ij} (v^{ik}) contains strictly positive (strictly negative) components. Note, that the properness of the contract v makes it possible (in principal) to rescind this contract by each member $i \in S(\mathrm{v})$ simply by not delivering the corresponding commodity to the participant $k = k(i)$.

Any finite set $\mathcal{V} = \{\mathrm{v}_r\}_{\mathcal{R}}$, consisting of proper contracts v_r, is called a (proper) *contract system of type* M_σ (*contractual* M_σ*-system*, or c.s. (of type M_σ), for short). Let us stress at once that the elements of $\mathcal{R}$ are supposed to be the "titles" of any type (not necessary natural numbers), naming the contracts belonging to the set $\mathcal{V}$.

Thus, it is supposed that the members of any coalition $S \in \sigma$ may enter several contracts. Moreover, a c.s. $\mathcal{V}$ may contain several samples of the same contract $\mathrm{v}(S)$ differing just by their titles (or, by their names, to be exact). To simplify the terms, we call a c.s. $\tilde{\mathcal{V}} = \{\tilde{\mathrm{v}}_r\}_{\tilde{\mathcal{R}}}$ *a subsystem of* c.s. $\mathcal{V} = \{\mathrm{v}_r\}_{\mathcal{R}}$ if and only if $\tilde{\mathcal{R}} \subseteq \mathcal{R}$, and $\tilde{\mathrm{v}}_r = \mathrm{v}_r$ for any $r \in \tilde{\mathcal{R}}$. Hence, we consider contract systems $\mathcal{V}$ and $\tilde{\mathcal{V}}$ to be identical if and only if $\mathcal{R} = \tilde{\mathcal{R}}$ and $\tilde{\mathrm{v}}_r = \mathrm{v}_r$ for any $r \in \mathcal{R}$.

Denote by $\Delta^i(\mathcal{V})$ the net outcome of the agent i resulting after a c.s. $\mathcal{V} = \{\mathrm{v}_r\}_{\mathcal{R}}$ is accepted and realized

$$\Delta^i(\mathcal{V}) := \begin{cases} \sum\limits_{r \mid i \in S(\mathrm{v}_r)} \sum\limits_{j \in S(\mathrm{v}_r)} \mathrm{v}_r^{ij}, & i \in \bigcup\limits_{r \in \mathcal{R}} S(\mathrm{v}_r), \\ 0 & \text{otherwise.} \end{cases}$$

Since $\mathrm{v}_r^{ij} = -\mathrm{v}_r^{ji}$ for any $i,j \in S(\mathrm{v}_r)$, $r \in \mathcal{R}$, we have that $\sum\limits_N \Delta^i(\mathcal{V}) = 0$ and, hence, $\sum\limits_N x^i(\mathcal{V}) = \sum\limits_N w^i$ with

$$x^i(\mathcal{V}) = x^i(\mathcal{V}, \mathcal{E}) := w^i + \Delta^i(\mathcal{V})$$

to be the resulting outcome of the agent i, obtained by means of the entering into c.s. $\mathcal{V}$. In the sequel we call $x(\mathcal{V}) = x(\mathcal{V}, \mathcal{E}) := (x^i(\mathcal{V}, \mathcal{E}))_N$ *a contractual* M_σ*-allocation* of economy $\mathcal{E}$ (c.a. (of type M_σ), for short).

Definition 2. *A contractual system* $\mathcal{V}$ *is called feasible, if* $x^i(\mathcal{V}) \in X_i$ *for any* $i \in N$.

By definition, feasibility of a c.s. $\mathcal{V}$ guarantees the contractual M_σ-allocation $x(\mathcal{V})$ to belong to the set

$$X(N) = X_{\mathcal{E}}(N) := \{(x^i)_N \in \prod_N X_i \mid \sum_N x^i = \sum_N w^i\}$$

of balanced allocations of the economy $\mathcal{E}$.

Remind, that due to the properness, any contract $\mathrm{v} \in \mathcal{V}$ may, in principal, be broken by any participant $i \in S(\mathrm{v})$. A formal description of the outcomes of rescinding (breaking) contractual subsystems of a feasible c.s. $\mathcal{V} = \{\mathrm{v}_r\}_{r\in\mathcal{R}}$ takes a bit more space. To start with, fix a subsystem $\mathcal{V}' = \{\mathrm{v}_r\}_{r\in\mathcal{R}'}$ of $\mathcal{V}$, consisting of the contracts to be broken, and put

$$\mathcal{U} := \mathcal{V} \setminus \mathcal{V}' = \{\mathrm{v}_r\}_{r\in\mathcal{T}}$$

with $\mathcal{T} := \mathcal{R} \setminus \mathcal{R}'$. Further, let $\mathcal{F}(\mathcal{V}, \mathcal{V}')$ be a collection of all the feasible contractual systems $\tilde{\mathcal{V}}$ of the economy $\mathcal{E}$, satisfying requirements:

(*) $\tilde{\mathcal{V}} = \{\mathrm{v}_r\}_{r\in\tilde{\mathcal{R}}}$ is a subsystem of $\mathcal{V}$.

Introduce, finally, a version $A_*(\mathcal{V}, \mathcal{R}', \mathcal{E})$ of the set of the breaking process outcomes, investigating in this paper:

$$A_*(\mathcal{V}, \mathcal{R}', \mathcal{E}) := \begin{cases} \{\mathcal{U}\} , & \text{if } \mathcal{U} \text{ is a feasible c.s. of } \mathcal{E}, \\ \mathcal{F}(\mathcal{V}, \mathcal{V}') & \text{otherwise.} \end{cases}$$

Thus, $A_*(\mathcal{V}, \mathcal{R}', \mathcal{E})$ contains all the outcomes of the breaking (cancelation) procedure, consisting of no more than two steps. At the first step of this procedure we nullify all the contracts v_r, $r \in \mathcal{R}'$. In case $\mathcal{U} = \{\mathrm{v}_r\}_{\mathcal{R}\setminus\mathcal{R}'}$ is a feasible subsystem, we stop the breaking procedure and put $A_*(\mathcal{V}, \mathcal{R}', \mathcal{E}) := \{\mathcal{U}\}$. Otherwise, at the second step, we continue to nullify contracts v_r, $r \in \mathcal{R}''$, with $\mathcal{R}''$ being a subset of $\mathcal{R} \setminus \mathcal{R}'$, in order to guarantee the feasibility of subsystem $\tilde{\mathcal{U}} = \{\mathrm{v}_r\}_{\mathcal{R}\setminus(\mathcal{R}'\cup\mathcal{R}'')}$. So, according to the requirement (*), any subsystem $\tilde{\mathcal{U}}$ of $\mathcal{U}$ can be chosen as a final outcome of the second step, provided that this subsystem is feasible. Certainly, there may be no outcomes at the second step in case $\mathcal{U}$ has no feasible subsystems at all.

Summarizing, we stress once more, that a breaking procedure at the second step (admissible when $\mathcal{U}$ is not feasible, only) supposed to be a spontaneous one. Hence, any feasible subsystem of $\mathcal{U}$ (if exists) may be realized as an outcome of this procedure. Thus, except when U is a feasible system itself $(A_*(\mathcal{V},\mathcal{R}',\mathcal{E})=\{\mathcal{U}\})$, we deal with the multiplicity of outcomes, in general. It is clear, even from the first glance, that the multiplicity mentioned causes significant obstacles in the study of contractual stability (for more details see Section 2 below).

To describe which way a coalition $S\in\sigma$ can improve a c.a. $x(\mathcal{V})$, generated by a feasible c.s. $\mathcal{V}=\{\mathrm{v}_r\}_{\mathcal{R}}$ of type M_σ, we suppose first that together with possibility to cancel some contracts v_r with $r\in\mathcal{R}^S=\mathcal{R}^S_{\mathcal{V}}:=\{r\in\mathcal{R}\mid S(\mathrm{v}_r)\bigcap S\neq\emptyset\}$ it is allowed for the coalition S to enter some new contract w. Denote by $\mathcal{E}_{\mathrm{w}}$ the modification of $\mathcal{E}$ generated by w:

$$\mathcal{E}_{\mathrm{w}}=\langle N,\{X_i,w^i+\Delta^i(\mathrm{w}),\alpha_i\}_N,\sigma\rangle,$$

where, as before, $\Delta^i(\mathrm{w}):=\Delta^i(\{\mathrm{w}\})$, i.e.,

$$\Delta^i(\mathrm{w})=\begin{cases}\sum\limits_{j\in\in S(\mathrm{w})}\mathrm{w}^{ij}, & \text{if } i\in S(\mathrm{w}),\\ 0 & \text{otherwise.}\end{cases}$$

Definition 3. *We say that a coalition $S\in\sigma$ with $|S|\geq 2$ w-improves (w-blocks) a feasible contractual M_σ-system $\mathcal{V}=\{\mathrm{v}_r\}_{\mathcal{R}}$, if there exists a subset $\mathcal{R}'\subseteq\mathcal{R}^S$, and a proper contract* w *of type M_S such that $S(\mathrm{w})=\mathrm{S}$, and for any $\mathcal{W}\in A_*(\mathcal{V},\mathcal{R}',\mathcal{E}_{\mathrm{w}})$ it holds: $x^i(\mathcal{W},\mathcal{E}_{\mathrm{w}})\in\alpha_i(x^i(\mathcal{V},\mathcal{E}))$ for any $i\in S$, with $x^i(\mathcal{W},\mathcal{E}_{\mathrm{w}})\in\mathcal{P}_i(x^i(\mathcal{V},\mathcal{E}))$ for at least one $i\in S$.*

As to the case of one-element coalition, $S=\{i\}\in\sigma$, it is said that S w-improves a feasible contractual M_σ-system $\mathcal{V}=\{\mathrm{v}_r\}_{\mathcal{R}}$, if there exists a subset $\mathcal{R}'\subseteq\mathcal{R}^S$ such that for any $\mathcal{W}\in A_*(\mathcal{V},\mathcal{R}',\mathcal{E})$ it holds: $x^i(\mathcal{W},\mathcal{E})$ belongs to $\mathcal{P}_i(x^i(\mathcal{V},\mathcal{E}))$.

(Here and below, as before, $\mathcal{P}_i(z):=\{x^i\in\alpha_i(z)\mid(x^i,z)\notin\alpha_i\}$ is the set of those bundles from X_i, which are strictly preferred to z.)

Definition 4. *A feasible contractual M_σ-system $\mathcal{V}$ is called a weak quasi-stable, if there is no coalition $S\in\sigma$, which w-improves $\mathcal{V}$.*

For any $x\in X(N)$ denote by $\mathcal{C}(x)=\mathcal{C}_{M_\sigma}(x)$ the set of all feasible c.s. $\mathcal{V}$ of type M_σ such that $x=x(\mathcal{V})$. To isolate the allocations $x\in X(N)$ with $\mathcal{C}_{M_\sigma}(x)\neq$

$\emptyset$, whose coalitional stability is independent on the concrete representation via some c.s. $\mathcal{V}$ of type M_σ, we introduce the main notion of the paper.

Definition 5. *An allocation $x \in X(N)$ is called a weak totally contractual M_σ-allocation (w.t.c.a. (of type M_σ), for short), if $C_{M_\sigma}(x)$ is nonempty set, and it contains only weak quasi-stable contractual systems (i.e., for every $\mathcal{V} \in C_{M_\sigma}(x)$ it holds: $\mathcal{V}$ is a weak quasi-stable allocation).*

A set $D_*^{M_\sigma}(\mathcal{E})$ of all the weak totally contractual M_σ-allocations of $\mathcal{E}$ is called *a weak totally contractual core (of type M_σ) of economy $\mathcal{E}$.*

Remark 1. Note, that like in this paper, the main point at issue in ([9]) is the fact that after by some coalition $S \in \sigma$ chosen contracts with numbers $r \in \mathcal{R}'$ are broken, the rest, with numbers $r \in \mathcal{R} \setminus \mathcal{R}'$ may not constitute a feasible contractual system in $\mathcal{E}_{\mathrm{W}}$. In both papers, it is supposed that the breaking process proceeds spontaneously, and stops after feasibility of the contractual system $\mathcal{U} = \{\mathrm{v}_r\}_{r\in\mathcal{R}\setminus\mathcal{R}'}$ is recovered. One of the distinctive feature of the stopping rule applied in the paper mentioned is that the set of outcomes (denoted there by $A(\mathcal{V},\mathcal{R}',\mathcal{E}_{\mathrm{W}})$) supposed to consist of only the maximal feasible subsystems of $\mathcal{U}$ (remind, that here, in this paper, we deal with $A_*(\mathcal{V},\mathcal{R}',\mathcal{E}_{\mathrm{W}})$ that consists of all the feasible subsystem of $\mathcal{U}$). Consequently, we have

$$A(\mathcal{V},\mathcal{R}',\mathcal{E}_{\mathrm{W}}) \subseteq A_*(\mathcal{V},\mathcal{R}',\mathcal{E}_{\mathrm{W}}).$$

Second (and the last) distinction concerns the blocking rule: to improve a feasible contractual M-system $\mathcal{V}$ it is sufficient (according to ([9])) to find at least one maximal feasible subsystem of $\mathcal{U}$ that improves $\mathcal{V}$ (unlike the w-blocking in this paper, which requires any feasible subsystem of $\mathcal{U}$ to be able to improve $\mathcal{V}$).
Summarizing, we have that the sets $D^M(\mathcal{E})$, $M \in \mathcal{M}$, of totally contractual M-allocations (unblocked M-allocation in the sense of blocking from ([9])) satisfy inclusions

$$D^M(\mathcal{E}) \subseteq D_*^M(\mathcal{E}), \quad M \in \mathcal{M}.$$

Surprisingly, for several rather wide classes of pure exchange economies it holds: $D^M(\mathcal{E}) = D_*^M(\mathcal{E})$ (for more details, see Sect.4 below).

Note, that any coalitional structure σ admits the (unique) representation as the union of (locally) irreducible coalitional substructures, inscribed into the corresponding components of N. Therefore, in what follows, we suppose σ to be irreducible itself. For the sake of completeness, remind corresponding definition from ([7]) (as usual, below we call a partition $\{N_1, N_2\}$ nontrivial, if N_1 and N_2 are nonempty):

- *Irreducibility-assumption*: for any nontrivial partition $\{N_1, N_2\}$ of N there is $S \in \sigma$ such that $S \bigcap N_1$ and $S \bigcap N_2$ are nonempty sets.

Besides, everywhere below it is assumed that all the subsets M_S are identical and equal to some linear subspace $M \subseteq \mathbf{R}^l$ satisfying the following *Sign-assumption*:

- *Sign-assumption*: every nonzero vector $z \in M$ contains both strictly positive and strictly negative components.

Remark 2. Observe, that from the Bipolar Theorem and Minkovski Separation Theorem it follows immediately that a linear subspace $M \neq \mathbf{R}^l$ satisfies the above mentioned Sign-assumption if and only if its polar subspace $M^0 := \{p \in \mathbf{R}^l \mid p \cdot z = 0,\ z \in M\}$ meets the requirement

$$M^0 \bigcap \mathrm{int}\mathbf{R}^l_+ \neq \emptyset. \tag{2}$$

It is clear, that in case relation (2) is satisfied there exists a nonempty finite subset $P_M \subseteq \mathbf{R}^l$ containing at least one strictly positive vector, such that $M = \{z \in \mathbf{R}^l \mid p \cdot z = 0,\ p \in P_M\}$. Hence, the economical meaning of the constraint imposed on the type of the contracts we consider below is as follows: elementary exchanges bundles v^{ij} should be compatible with all the fixed-price vectors p, belonging to some nonempty finite subset $P_M \subseteq \mathbf{R}^l$, containing at least one strictly positive price vector $\overline{p}$.
Under Sign-assumption it can easily be shown also , that for any $S \subseteq N$ with $|S| \geq 3$ there exists a proper "zero-contract" w of type M (i.e., a proper contract $\mathrm{w} \neq 0$ of type M such that (i)$S(\mathrm{w}) = S$, and (ii)$\Delta^i(\mathrm{w}) = 0$ for any $i \in S$). It means that the properness of any contract v of type $M \neq \{0\}$ with $|S(\mathrm{v})| \geq 3$ is guaranteed "automatically": $\mathrm{u} := \mathrm{v} + \lambda \mathrm{w}$ is a proper contract of type M with $S(\mathrm{u}) = S(\mathrm{v})$ and $\Delta^i(\mathrm{u}) = \Delta^i(\mathrm{v}),\ i \in N$, provided that $\mathrm{w} \neq 0$ is a proper "zero-contract" of type M with $S(\mathrm{w}) = S(\mathrm{v})$, and $\lambda > 0$ is large enough. Hence, everywhere below we may (and will) deal with some contracts and contract systems of type M, not necessarily satisfying properness-assumption for not-two-person

coalitions (at least, in those situations, where only properness "modulo zero-contract" matters).

Denote by $\mathcal{C}_M$ a collection of all c.s. of type M_σ with $M_S = M$ for any $S \in \sigma$. In what follows, the contracts v of type $M_S = M$, as well as the systems $\mathcal{V} \in \mathcal{C}_M$, and allocations $x = x(\mathcal{V})$, will be called *M-contracts, M-systems and M-allocations*, respectively. To present more detailed description of some unblocked (in the sense of Definition 3.) M-allocations $x(\mathcal{V})$, we characterize first all feasible allocations, attainable by entering into the contract systems of type M. Put

$$X_{\mathcal{E}}^{M} := \{x \in \prod_N X_i \mid \exists \mathcal{V} \in \mathcal{C}_M : x = x(\mathcal{V})\}.$$

Proposition 1. below was established in ([8]); for its crucial role in the further considerations and for the sake of completeness we reproduce it here together with the proof that seems to be rather instructive.

Proposition 1. *If* σ *is irreducible coalition structure, then*

$$X_{\mathcal{E}}^{M} = \{x \in X(N) \mid \exists \Delta \in \Delta_M(N) : x = w + \Delta\},$$

where $\Delta_M(N) := \{\Delta = (\Delta^i)_N \in M^N \mid \sum_N \Delta^i = 0\}$, $w := (w^i)_N$.

Proof. If $x \in X_{\mathcal{E}}^{M}$, then $x = x(\mathcal{V}) = w + \Delta(\mathcal{V})$ for some $\mathcal{V} \in \mathcal{C}_M$, where $\Delta(\mathcal{V}) := (\Delta^i(\mathcal{V}))_N$. By definition of M-contract it follows that $\Delta^i(\mathcal{V}) \in M$, $\sum_N \Delta^i(\mathcal{V}) = 0$ and, consequently, we get desired: $x = w + \Delta$ for some $\Delta \in \Delta_M(N)$.

Let now $x \in X(N)$ with $x = w + \Delta$ for some $\Delta \in \Delta_M(N)$. To prove that there is $\mathcal{V} \in \mathcal{C}_M$, such that $\Delta = \Delta(\mathcal{V})$, we apply induction on the number $|N|$ of the economical agents of $\mathcal{E}$. It is clear, that in case $|N| = 2$ we have : $\mathcal{V} = \{\text{v}\}$, with $\text{v}^{12} := \Delta^1$, and $\text{v}^{21} := \Delta^2$. Suppose that our assertion is valid for all economies of type (1) with $|N| \leq m$. Fix some economy $\mathcal{E}$ with $|N| = |N_{\mathcal{E}}| = m+1$. If $N \in \sigma$, then, modulo "zero-contract", in accord with Remark 2, a c.s. required consists of the only M-contract v, defined by the formula: $\text{v}^{ii+1} := \sum_{k=1}^{i} \Delta^k$, $i = 1, \ldots, m$; $\text{v}^{ij} = 0$, $j - i > 1$ (note, that due to the equalities $\text{v}^{ij} = -\text{v}^{ji}$ to give a complete description of the contract v it is sufficient to indicate all the vectors v^{ij} with $i, j \in S(\text{v})$ satisfying inequalities $i < j$).

Consider the case $N \notin \sigma$. Entering into the trivial contracts for some coalitions $S \in \sigma$, if necessary, we may assume without loss of generality, that σ is minimal (w.r.t. inclusion in 2^N) irreducible coalitional structure. Fix some $S^0 \in \sigma$. It is not hard to prove that there exists a partition $\eta^0 = \{N_1^0, N_2^0\}$ of N such that (i)coalitions $S_k^0 := S^0 \bigcap N_k^0$, $k = 1,2$, are nonempty; (ii)for any $T \in \sigma \setminus \{S^0\}$ either $T \subseteq N_1^0$, or $T \subseteq N_2^0$; and (iii)coalitional structures $\sigma_k^0 := \{T \in \sigma \mid T \subseteq N_k^0\} \bigcup \{S_k^0\}$, $k = 1,2$, are irreducible in N_1^0, N_2^0, respectively. Indeed, existence of a partition η^0, satisfying conditions (i) and (ii), follows immediately from the minimality of the irreducible coalitional structure σ (since in case $\sigma \setminus \{T\}$ is irreducible for some $T \in \sigma$ we get a contradiction). As to the irreducibility of σ_1^0, for example, it can easily be shown by consideration of three cases for any nontrivial partition $\{N_1^1, N_1^2\}$ of N_1^0:

1) $S^0 \bigcap N_1^k \neq \emptyset, k = 1,2;$

2) $S^0 \bigcap N_1^1 \neq \emptyset, S^0 \bigcap N_1^2 = \emptyset;$

3) $S^0 \bigcap N_1^1 = \emptyset, S^0 \bigcap N_1^2 \neq \emptyset.$

Specifically, considering partitions $\{N_1^2, N_1^1 \bigcup N_2\}$ and $\{N_1^1, N_1^2 \bigcup N_2\}$ in cases 2) and 3), respectively, we get required coalitions from σ_1^0 on the basis of condition (ii). In fact, since in both these cases coalition S^0 doesn't intersect one of the elements of the corresponding nontrivial partition of N, by (ii) and irreducibility of σ, in each of these cases there exists a coalition $S \in \sigma_1^0$ such that $S \bigcap N_1^k \neq \emptyset$ for any $k = 1,2$.

Now, by making use of the partition η^0, we divide initial economy $\mathcal{E}$ into two appropriate "smaller" economies

$$\mathcal{E}_{(k)} := \langle N_{(k)}, X_i^{(k)}, w_{(k)}^i, \alpha_i^{(k)} \}_{N_{(k)}}, \sigma_{(k)} \rangle, \quad k = 1,2,$$

satisfying inductive hypothesis and "implementing" the allocation $x = w + \Delta$ with $\Delta = (\Delta^1, \ldots, \Delta^n) \in \Delta_M(N)$ by means of some contractual M-system (as a result of a suitable exchange, both within and between these economies). To this end we fix some economic agents $i_k \in S_k^0$, $k = 1,2$, and put

$$N_{(k)} := N_k^0, \quad k = 1,2;$$

$$w_{(k)}^i := w^i, \quad i \in N_{(k)}, k = 1,2;$$

$$X_i^{(k)} := X_i \quad i \in N_{(k)} \setminus \{i_k\}, k = 1,2.$$

As to the consumption sets $X_{i_1}^{(1)}$ and $X_{i_2}^{(2)}$, we put $X_{i_1}^{(1)} := \{w^{i_1} - \sum_{i \in N_{(1)} \setminus \{i_1\}} \Delta^i\}$ and $X_{i_2}^{(2)} := \{w^{i_2} + \sum_{i \in N_{(1)} \cup \{i_2\}} \Delta^i\}$. Finally, we take $\sigma_{(k)}$ to be equal to σ_k^0, $k = 1,2$, and put $\alpha_i^{(k)} := \alpha_i$ for any $i \in N_{(k)}$ and $k = 1,2$. To apply inductive hypothesis, take $\Delta_{(k)} \in \Delta_M(N_{(k)})$, $k = 1,2$, with

$$\Delta_{(1)}^{i_1} := - \sum_{i \in N_{(1)} \setminus \{i_1\}} \Delta^i, \ \Delta_{(2)}^{i_2} := \sum_{i \in N_{(1)} \cup \{i_2\}} \Delta^i,$$

and $\Delta_{(k)}^i := \Delta^i$ for any $i \in N_{(k)} \setminus \{i_k\}$, $k = 1,2$. Further, for each $k = 1,2$, put $w_{(k)} := (w^i)_{N_{(k)}}$ and $\Delta_{(k)} := (\Delta_{(k)}^i)_{N_{(k)}}$. It is clear, that $x_{(1)} := w_{(1)} + \Delta_{(1)}$ and $x_{(2)} := w_{(2)} + \Delta_{(2)}$ are balanced allocations of the economies $\mathcal{E}_{(1)}$, $\mathcal{E}_{(2)}$, respectively. Hence, considerations, given above, implies: economies $\mathcal{E}_{(1)}$, $\mathcal{E}_{(2)}$ and corresponding allocations $x_{(1)}$, $x_{(2)}$ satisfy our inductive hypothesis, and, consequently, there exist M-systems $\mathcal{V}_{(1)}$ and $\mathcal{V}_{(2)}$ such that $x_{(k)} = x(\mathcal{V}_{(k)}, \mathcal{E}_{(k)})$, each $k = 1,2$.

Now we design a contract $v_{(S^0)}$, which coalition S^0 should enter to in order to organize (together with the M-systems $\mathcal{V}_{(1)}$ and $\mathcal{V}_{(2)}$) an M-system $\mathcal{V}$, satisfying equality $x = x(\mathcal{V}) = x(\mathcal{V}, \mathcal{E})$. To do so, we put

$$v_{(S^0)}^{i_1 i_2} := \sum_{i \in N_{(1)}} \Delta^i, \quad v_{(S^0)}^{i_2 i_1} := - \sum_{i \in N_{(1)}} \Delta^i,$$

and $v_{(S^0)}^{ij} := 0$ for any $i,j \in S^0$ such that $\{i,j\} \neq \{i_1, i_2\}$. It is clear, that the system $\mathcal{V} := \mathcal{V}_{(1)} \bigcup \mathcal{V}_{(2)} \bigcup \{v_{(S^0)}\}$, with $v_{(S^0)}$ thus defined (and modified in the spirit of of Remark 2, if necessary), meets our requirement, which proves the proposition. Q.E.D.

3. Weak Totally Contractual Set and Equilibrium Allocations

Below, we restrict ourselves to the case, when M is a hypersubspace of $\mathbf{R}^l$ (i.e., we assume, that $\dim M = l - 1$). In this situation it turns out that any w.t.c. allocation is an equilibrium allocation under fairly natural regularity assumptions. Vice versa, under essentially weaker assumptions any (competitive) equilibrium allocation turns out to be a w.t.c. allocation, provided that a subspace M is properly chosen.

Remind (see, e.g., [1], [3]), that $\bar{x} \in X(N)$ is called a *competitive equilibrium allocation* (equilibrium allocation, for short), if there exists a nonzero vector $\bar{p} \in \mathbf{R}^l$ such that

$$\mathcal{P}_i(\bar{x}^i) \bigcap B_i(\bar{p}) = \emptyset, \quad i \in N, \tag{3}$$

where, as usual, $B_i(\bar{p})$ is the budget set of an agent i at prices $\bar{p}$:

$$B_i(\bar{p}) := \{x^i \in X_i \mid \bar{p} \cdot x^i \leq \bar{p} \cdot w^i\}.$$

Denote by $W(\mathcal{E})$ the set of all equilibrium allocations of economy $\mathcal{E}$. Remind, that any vector $\bar{p}$, satisfying (3), is called an equilibrium price vector (equilibrium price, for short), supporting the allocation $\bar{x}$.

Theorem 1. *Let $\bar{x}$ be an equilibrium allocation of the economy $\mathcal{E}$. If there is a strictly positive equilibrium price vector $\bar{p}$, supporting $\bar{x}$, then $\bar{x}$ belongs to $D_*^M(\mathcal{E})$ with $M := \{z \in \mathbf{R}^l \mid \bar{p} \cdot z = 0\}$.*

Proof. Since $\bar{x} \in W(\mathcal{E})$ implies inclusion $\bar{x} \in X(N) \bigcap \prod_N B_i(\bar{p})$, we have: $\Delta^i := \bar{x}^i - w^i$ belongs to M for any $i \in N$. It is clear, that irreducibility of σ and Proposition1. imply: $\bar{x} = w + \Delta \in X_{\mathcal{E}}^M$, where, as before, $w = (w^i)_N$, and $\Delta = (\Delta^i)_N$. Suppose, the allocation $\bar{x}$ can be w-improved by some coalition $S \in \sigma$. Then, by Definition 3., there exist at least one contractual system $\mathcal{V} \in C_M(\bar{x})$ and one participant $i \in S$ such that $x^i(\mathcal{V}) \in \mathcal{P}_i(\bar{x}^i)$ and $x^j(\mathcal{V}) \in \alpha_j(\bar{x}^j)$ for any $j \in S \setminus \{i\}$. By definition of equilibrium allocation, the former inclusion $x^i(\mathcal{V}) \in \mathcal{P}_i(\bar{x}^i)$ implies $\mathcal{P}_i(\bar{x}^i) \bigcap B_i(\bar{p}) = \emptyset$. But then $\bar{p} \cdot x^i(\mathcal{V}) > \bar{p} \cdot w^i$, which contradicts to the inclusion $x^i(\mathcal{V}) \in X_i^M$ yielding $\bar{p} \cdot x^i(\mathcal{V}) = \bar{p} \cdot w^i$.

Thus, there are no coalitions $S \in \sigma$ that can w-improve $\bar{x}$ and, hence $\bar{x}$ belongs to the weak totally contractual core (of type M) of the economy $\mathcal{E}$. Q.E.D.

To prove the reverse inclusion $D_*^M(\mathcal{E}) \subseteq W(\mathcal{E})$ we have to add some compatibility assumptions concerning the coalitional structure σ (as well, as traditional convexity and monotonicity assumptions, imposed on the consumption sets X_i and individual preference relations α_i). Namely, from now on it is supposed that for any $i \in N$ it holds:

(a) X_i is a convex set;

(b) α_i is a reflexive binary relation;

(c) for any $x^i \in \widehat{X}_i := \mathrm{Pr}_{X_i} X(N)$ there exists $z \in \mathrm{int}\mathbf{R}^l_+$ such that $x^i + z \in \mathcal{P}_i(x^i)$;

(d) for any $x^i \in X_i$ the set $\mathcal{P}_i(x^i)$ is convex and, besides, $(x^i, y^i] \subseteq \mathcal{P}_i(x^i)$ for any $y^i \in \mathcal{P}_i(x^i)$, where $(x^i, y^i] := \{(1-t)x^i + ty^i \mid t \in (0,1]\}$.

To propose a compatibility assumption concerning σ, remind first, that for any $i \in N$ we denote by σ_i a set of those coalitions $S \in \sigma$ that contain i :

$$\sigma_i := \{S \in \sigma \mid i \in S\}.$$

Second, we introduce a useful strengthening of the notion of σ-divisibility, proposed earlier in ([7]).

Definition 6. *A coalition $T \subseteq N$ is said to be strongly σ-divisible, if for any $i \in N$ there are two coalitions $R, S \in \sigma_i$ such that*

$$R \cap (T \setminus S) \neq \emptyset. \tag{4}$$

Further, as in ([9]), for any $x \in X^M_{\mathcal{E}}$ put

$$N_x := \{i \in N \mid x^i \in \mathrm{int}_M X_i\},$$

where $\mathrm{int}_M X_i$ is the (relative) interior of $X_i^M := (w^i + M) \cap X_i$ in the affine subspace $w^i + M$.

The main result of this section is as follows.

Theorem 2. *Suppose $\bar{x}$ belongs to $D^M_*(\mathcal{E})$ for some M satisfying Sign-assumption. If $N_{\bar{x}}$ is a strongly σ-divisible subset of N, then $\bar{x}$ is an equilibrium allocation of $\mathcal{E}$.*

Proof. Let $\bar{x} \in D^M_*(\mathcal{E})$, and i be any agent of the economy $\mathcal{E}$. In order to show that $\bar{x}^i$ is a locally maximal element on X_i^M w.r.t. the binary relation α_i, we fix first some coalitions $R, S \in \sigma_i$ and a participant $k \in R$ such that $k \in N_{\bar{x}}$ and $k \notin S$ (remind, that the existence of such coalitions R, S and a participant k follows directly from the fact that $N_{\bar{x}}$ is a strongly σ-divisible set). Since $\bar{x}^k \in \mathrm{int}_M X_k$, there is some $U \subseteq M$, being a symmetric neighbourhood of zero in M, such that $\bar{x}^k + U \subseteq X_k^M$. Assuming that $\bar{x}^i$ is not a locally maximal element on X_i^M w.r.t. the binary relation α_i, we have that there exists $z \in U$ satisfying inclusion $\bar{x}^i + z \in \mathcal{P}_i(\bar{x}^i)$. To get a contradiction, we construct now a contractual

M-system $\overline{\mathcal{V}} \in \mathcal{C}(\overline{x})$, which can be w-improved by coalition S (chosen in the very beginning of the proof). To do so, fix zero M-contract w with $S(\mathrm{w}) = \mathrm{S}$ (and $\mathrm{v}^{ij} = 0$ for any $\{i,j\} \subseteq S$), and consider an arbitrary $\mathcal{V} \in \mathcal{C}(\overline{x})$. Further, denote by $\overline{\mathrm{v}}$ an M-contract satisfying requirements: $S(\overline{\mathrm{v}}) = R$, and

$$\overline{\mathrm{v}}^{jr} = \begin{cases} z \ , & j = i, r = k, \\ 0 \ , & \{j,r\} \neq \{i,k\} \end{cases}$$

(it follows directly from Remark 2 that it always possible to construct a proper "almost zero-contract" v of type M with $|S(\mathrm{v})| \geq 3$ and $\mathrm{v}^{ij} = 0$ for all but one two-element subsets $\{i,j\} \subseteq S$). Put $\mathcal{V} := \mathcal{V} \cup \{\mathrm{v}_1, \mathrm{v}_2\}$ with $\mathrm{v}_1 = \overline{\mathrm{v}}$, $\mathrm{v}_2 = -\overline{\mathrm{v}}$, where, as usual, $(-\mathrm{v})^{ij} = -\mathrm{v}^{ij}$ for any $\{i,j\} \subseteq S(\mathrm{v})$. Further, consider $\mathcal{R}' \subseteq \mathcal{R}^S$ to be equal to $\{2\}$. Note first, that by definition of the contracts v, w and contractual system $\mathcal{V}$ it follows immediately that $A_*(\mathcal{V}, \mathcal{R}', \mathcal{E}_\mathrm{w})$ contains only one system, namely, $A_*(\mathcal{V}, \mathcal{R}', \mathcal{E}_\mathrm{w}) = \{\mathcal{U}\}$ with $\mathcal{U} := \mathcal{V} \cup \{\mathrm{v}_1\}$ (no multiplicity of outcomes of cancelation process due to a proper chosen breaking system $\mathcal{R}'$). Second, due to the fact that preference relations α_j, $j \in N$, are reflexive, and by definition of z and v_1 we get: $x^j(\mathcal{U}, \mathcal{E}_\mathrm{w}) \in \alpha_j(x^j(\mathcal{V}, \mathcal{E}))$ for any $j \in S$, with $x^i(\mathcal{U}, \mathcal{E}_\mathrm{w}) \in \mathcal{P}_i(x^i(\mathcal{V}, \mathcal{E}))$. Hence, according to the Definition 3., we conclude, that coalition S w-improves a feasible contractual M-system $\mathcal{V}$ belonging to $\mathcal{C}(\overline{x})$.

Thus, we have obtained desired contradiction with assumption $\overline{x} \in D_*^M(\mathcal{E})$ and, consequently, finished the proof of the fact that for any $i \in N$, the bundle $\overline{x}^i$ is a local maximum on X_i^M w.r.t. his preference relation α_i.

More precisely, we have established that for any $i \in N$, there exists a symmetric neighbourhood of zero in M, say V, such that $\mathcal{P}_i(\overline{x}^i) \cap (\overline{x}^i + V) \cap X_i = \emptyset$. To continue the proof of the inclusion $\overline{x} \in W(\mathcal{E})$, check first that in fact the intersection $\mathcal{P}_i(\overline{x}^i) \cap X_i^M$ is empty, as well. To verify the latter assertion, just mention, that for any $y^i \in \mathcal{P}_i(\overline{x}^i) \bigcap X_i^M$ (if such y^i exists) the bundle $\overline{z}(t) := ty^i + (1-t)\overline{x}^i$ belongs to the neighbourhood $\overline{x}^i + V$ of $\overline{x}^i$ for any $t \in (0,1)$ small enough. But this fact (together with the convexity assumption (d)) contradicts to the local maximality of $\overline{x}^i$, which had already been proven above.

Summarizing, we have that for every $i \in N$ the consumption bundle $\overline{x}^i$ is maximal on X_i^M w.r.t. α_i. Proceeding now like in [8] (Theorem 4.13), pick some $\overline{p} \in M^0 \bigcap \mathrm{int}\mathbf{R}_+^l$ and prove that $\mathcal{P}_i(\overline{x}^i) \bigcap B_i(\overline{p}) = \emptyset$ for any $i \in N$. Let $y^i \in \mathcal{P}_i(\overline{x}^i)$ for some $i \in N$. Since $\overline{p} \cdot x^i = \overline{p} \cdot w^i$ for any $x^i \in X_i^M$ (in particular, $\overline{p} \cdot \overline{x}^i = \overline{p} \cdot w^i$, each $i \in N$), from the maximality of $\overline{x}^i$ on X_i^M it follows that $\overline{p} \cdot y^i \neq \overline{p} \cdot w^i$. Suppose, that $\overline{p} \cdot y^i < \overline{p} \cdot w^i$. Because of the inclusion $\overline{x} \in X(N)$ we have, due to the monotonicity assumption (c): there is $z \in \mathrm{int}\mathbf{R}^l_+$ such that $\bar{y}^i := \bar{x}^i + z \in \mathcal{P}_i(\bar{x}^i)$. It is clear, that $\bar{p} \cdot \bar{y}^i > \bar{p} \cdot w^i$. But then there exists $\bar{t} \in (0,1)$ such that the bundle $z(\bar{t}) := \bar{t}y^i + (1-\bar{t})\bar{y}^i$ belongs to X_i^M. On the other hand, due to the convexity of $\mathcal{P}_i(\bar{x}^i)$ we have: $z(\bar{t}) \in \mathcal{P}_i(\bar{x}^i)$, which contradicts to the maximality of $\bar{x}^i$ on X_i^M w.r.t. α_i.

Contradiction obtained proves the inequality $\bar{p} \cdot y^i > \bar{p} \cdot w^i$, which implies the relation $y^i \notin B_i(\bar{p})$. Due to the arbitrariness of element y^i taken from $\mathcal{P}_i(\bar{x}^i)$, we have desired: $\mathcal{P}_i(\bar{x}^i) \cap B_i(\bar{p}) = \emptyset$. Q.E.D.

Note, that the cardinality of $N_{\bar{x}}$ may be very small. It is clear, however, that σ-divisibility condition takes the simplest form in case $N_{\bar{x}} = N$.

To easify formulations, from now on we assume that hypersubspace M under consideration always satisfies Sign-assumption.

Corollary 1. *If* $|\sigma_i| \geq 2$ *for any* $i \in N$, *then* $D^M_{**}(\mathcal{E}) \subseteq W(\mathcal{E})$, *where*

$$D^M_{**}(\mathcal{E}) := D^M_{*}(\mathcal{E}) \bigcap X^M_0, \quad X^M_0 := \prod_N int_M X_i.$$

Proof. Let $\bar{x} \in D^M_{**}(\mathcal{E})$ be given. To prove that $\bar{x}$ belongs to $W(\mathcal{E})$, we mention first that due to the inclusion $\bar{x} \in X^M_0$ we have $N_{\bar{x}} = N$. To show that $\bar{x}$ satisfies all the requirements of Theorem2., it is sufficient to prove that under assumptions of our corollary the grand coalition N is strongly σ-divisible. Fix an arbitrary $i \in N$. By applying assumption $|\sigma_i| \geq 2$, select two different coalitions R and S, belonging to σ_i. Due to $R \neq S$ we have that either $R \setminus S \neq \emptyset$, or $S \setminus R \neq \emptyset$ (otherwise we get equality $R = S$ contradicting the choice of R, S). It is clear, that in both cases the relation (4) is satisfied (with $T = N$ and renaming the coalitions selected, if necessary). Hence, $N_{\bar{x}}$ is a strongly σ-divisible subset of N and, consequently, by Theorem 2. we get inclusion required: $\bar{x} \in W(\mathcal{E})$. Q.E.D.

The results obtained allow us to find some rather large classes of exchange models, admitting an equilibrium characterization for any w.t.c. allocation $\bar{x}$, without direct analysis of the structure of $N_{\bar{x}}$. To give some examples, introduce first additional characteristics of the exchange model $\mathcal{E}$. Denote by $\mathcal{M}$ the set of all hypersubspaces $M \subseteq \mathbf{R}^l$ satisfying Sign-assumption, and for any $M \in \mathcal{M}$ put

$$S^M_{\mathcal{E}} := \{i \in N \mid \alpha_i \text{ is complete, transitive, } w^i \in X_i, \text{ and } \alpha_i(w^i) \bigcap X^M_i \subseteq \mathrm{int}_M X_i\}.$$

Definition 7. *An exchange model $\mathcal{E}$ is called C_M-regular, if $S_{\mathcal{E}}^M \neq \emptyset$.*

Definition 8. *An exchange model $\mathcal{E}$ is called $C_{\mathcal{M}}$-regular, if it is C_M-regular for any $M \in \mathcal{M}$.*

Remark 3. Note, that by definition of the sets $S_{\mathcal{E}}^M$ we have that for any $M \in \mathcal{M}$ it holds: $w^i \in \mathrm{int}_M X_i$ for each $i \subset S_{\mathcal{E}}^M$. Consequently, for any $C_{\mathcal{M}}$-regular economy $\mathcal{E}$ we have: $w^i \in \mathrm{int}_M X_i$ for any $M \in \mathcal{M}$ and $i \in S_{\mathcal{E}}^M$.

By applying, mutatis mutandis, argumentation, used in the proof of Theorem 2., we obtain the following results.

Proposition 2. *Let $\mathcal{E}$ be a C_M-regular pure exchange model with w^i belonging to X_i for any $i \in N \setminus S_{\mathcal{E}}^M$. If $S_{\mathcal{E}}^M$ is a strongly σ-divisible subset of N, and one-element coalition $\{i\}$ belongs to σ for any $i \in S_{\mathcal{E}}^M$, then $D_*^M(\mathcal{E}) \subseteq W(\mathcal{E})$.*

Proof. Observe first, that by definition of the set $S_{\mathcal{E}}^M$ and due to the assumption $w^i \in X_i$, $i \in N \setminus S_{\mathcal{E}}^M$, of our proposition, we have that $w^i \in X_i$ for any $i \in N$ (initial endowments of the agents belong to their consumption sets). Let $\bar{x}$ be an arbitrary element of $D_*^M(\mathcal{E})$. It is not very hard to verify that $S_{\mathcal{E}}^M$ is contained in $N_{\bar{x}}$. To start with the proof of this assertion, note that due to the completeness of α_i and inclusion $\{i\} \in \sigma$, fulfilled for any $i \in S_{\mathcal{E}}^M$, we get: $\bar{x}^i \in \alpha_i(w^i)$ for any $i \in S_{\mathcal{E}}^M$. Indeed, suppose $\bar{x}^i \notin \alpha_i(w^i)$ for some $i \in S_{\mathcal{E}}^M$. Then, by completeness of α_i we get $w^i \in \mathcal{P}_i(\bar{x}^i)$. Applying the latter inclusion, we prove that the coalition $\{i\}$ can w-improve the allocation $\bar{x}$. Doing so, fix some $\mathcal{V} \in C(\bar{x})$ and put $S = \{i\}$, $\mathrm{w} = 0$ and $\mathcal{R}' = \mathcal{R}^S$. Note, that due to the inclusions $w^i \in X_i$, $i \in N$, we have that $A_*(\mathcal{V}, \mathcal{R}', \mathcal{E}_{\mathrm{w}}) \neq \emptyset$ (to argue, just mention, that in case the contractual system $\mathcal{U} := \{\mathrm{v_r}\}_{\mathrm{r} \in \mathcal{R} \setminus \mathcal{R}^S}$ happens not to be feasible, we can just break all the rest, nullifying all the contracts from $\mathcal{V} \setminus \mathcal{U}$; the resulting empty system gives us initial endowment allocation $w = (w^1, \ldots, w^n)$, which belongs to $X(N)$ due to the inclusions $w^i \in X_i$ mentioned). Moreover, it is easy to check that $x^i(\mathcal{V}', \mathcal{E}_{\mathrm{w}}) = w^i$ for any $\mathcal{V}' \in A_*(\mathcal{V}, \mathcal{R}', \mathcal{E}_{\mathrm{w}})$ and, hence, according to the Definition 3. and our supposition $w^i \in \mathcal{P}_i(\bar{x}^i)$ we have: coalition $S = \{i\}$ w-improves the allocation $\bar{x}$. But the latter contradicts to the assumption $\bar{x} \in D_*^M(\mathcal{E})$. Due to the contradiction obtained, we get: $\bar{x}^i$ belongs to $\alpha_i(w^i)$ for any $i \in S_{\mathcal{E}}^M$. Since, evidently, $\bar{x}^i$ belongs to X_i^M for any $i \in N$, by inclusions $\bar{x}^i \in \alpha_i(w^i)$ and $\alpha_i(w^i) \cap X_i^M \subseteq \mathrm{int}_M X_i$, $i \in S_{\mathcal{E}}^M$, we get: $\bar{x}^i \in \mathrm{int}_M X_i$ for any $i \in S_{\mathcal{E}}^M$. Thus, the inclusion

$$S_{\mathcal{E}}^M \subseteq N_{\bar{x}}, \tag{5}$$

mentioned in the beginning of our proof, is established.

It is clear, that any superset of a strongly σ-divisible set is a strongly σ-divisible set, as well. Hence, by inclusion (5) we have that $N_{\bar{x}}$ is a strongly σ-divisible set. Consequently, by applying Theorem 2 we get required: $\bar{x} \in W(\mathcal{E})$. Q.E.D.

Below, we demonstrate one of the quite clear implications of Proposition 2.

Corollary 2. *If $\mathcal{E}$ is C_M-regular exchange economy with $w^i \in X_i$ for any $i \in N \setminus S^M_{\mathcal{E}}$, one-element coalition $\{i\}$ belongs to σ for any $i \in N$, and, moreover, for any $i \in N$ there exists a coalition $R \in \sigma_i$ such that*

$$(R \setminus \{i\}) \cap S^M_{\mathcal{E}} \neq \emptyset, \tag{6}$$

then every weak totally contractual M-allocation is an equilibrium allocation.

Proof. The main lines of the proof are the same, as in the proof of Proposition 2. First, we establish that $\bar{x}^i$ belongs to $\alpha_i(w^i)$ for any $\bar{x} \in D^M_*(\mathcal{E})$ and $i \in S^M_{\mathcal{E}}$. Here we apply argumentation from the proof of Proposition 2., stating that assumption $w^i \in \mathcal{P}_i(\bar{x}^i)$ with $i \in S^M_{\mathcal{E}}$ implies that the one-element coalition $\{i\}$ w-improves allocation $\bar{x}$. Second, as a consequence of inclusions $\bar{x}^i \in \alpha_i(w^i)$, $i \in S^M_{\mathcal{E}}$, we get the insertion $S^M_{\mathcal{E}} \subseteq N_{\bar{x}}$. Finally, by taking for any $i \in N$ coalitions $S = \{i\}$ and R to be equal to that appearing in (6), we have that $S^M_{\mathcal{E}}$ is a strongly σ-divisible subset of N. Referring to Proposition 2., we complete the proof. Q.E.D.

To present a "global" version of the results, obtained in term of the C_M-regularity, we introduce one more important notion. Put

$$D_*(\mathcal{E}) := \bigcup_{M \in \mathcal{M}} D^M_*(\mathcal{E}), \tag{7}$$

$$S_{\mathcal{E}} := \bigcup_{M \in \mathcal{M}} S^M_{\mathcal{E}}. \tag{8}$$

Definition 9. *We call $D_*(\mathcal{E})$, defined by the formula (7), a weak totally contractual core of an economy $\mathcal{E}$.*

Theorem 3. *If $\mathcal{E}$ is C_M-regular, $S^M_{\mathcal{E}}$ is strongly σ-divisible for every $M \in \mathcal{M}$, $\{i\} \in \sigma$ for every $i \in S_{\mathcal{E}}$ (with $S_{\mathcal{E}}$ to be defined by (8)), and $w^i \in X_i$ for any $i \in$*

$N \setminus S_{\mathcal{E}}$, then the weak totally contractual core $D_*(\mathcal{E})$ of economy $\mathcal{E}$ is contained in the set $W(\mathcal{E})$ of competitive equilibria of $\mathcal{E}$.

In order to provide an easier verifiable regularity condition than that, given in Definition 8., we introduce another characteristic of the exchange model under consideration:

$$T_{\mathcal{E}} := \{i \in N \mid \alpha_i \text{ is complete, transitive, } w^i \in X_i, \text{ and } \alpha_i(w^i) \bigcap \check{X}_i \subseteq \text{int}X_i\},$$

where $\check{X}_i := X_i \setminus (w^i + \text{int}\mathbf{R}^l_+),\ i \in N$.

Definition 10. *An exchange model $\mathcal{E}$ is called C-regular, if $T_{\mathcal{E}} \neq \emptyset$.*

It is clear, that a slight modification of the proof of Proposition 2 gives the following analogs of Theorem 3.

Theorem 4. *If $T_{\mathcal{E}}$ is strongly σ-divisible, $w^i \in X_i$ for any $i \in N \setminus T_{\mathcal{E}}$, and $\{i\} \in \sigma$ for every $i \in T_{\mathcal{E}}$, then $D_*(\mathcal{E}) \subseteq W(\mathcal{E})$.*

Proof. Observe, that in the theorem under consideration C-regularity of $\mathcal{E}$ follows from the fact that it is assumed $T_{\mathcal{E}}$ to be a strongly σ-divisible subset of N and, hence, $T_{\mathcal{E}} \neq \emptyset$. Since, in some sense, the role of $T_{\mathcal{E}}$ in our theorem is quite similar to that, played by $S^M_{\mathcal{E}}$ in Proposition 2., we just adapt here the main lines of the proof of this proposition.

Let $\bar{x} \in D^M_*(\mathcal{E})$ for some $M \in \mathcal{M}$. Due to the assumption that $T_{\mathcal{E}}$ is strongly σ-divisible, the only thing we need to directly apply Theorem 2. and get the inclusion required $\bar{x} \in W(\mathcal{E})$, is to prove the insertion $T_{\mathcal{E}} \subseteq N_{\bar{x}}$. To this end, according to the assumptions $\alpha_i(w^i) \cap \check{X}_i \subseteq \text{int}X_i$, $i \in T_{\mathcal{E}}$, given, it would be enough to establish inclusions: $\bar{x}^i \in \check{X}_i$, and $\bar{x}^i \in \sigma_i(w^i)$ for any $i \in T_{\mathcal{E}}$. Note, that the first inclusions, $\bar{x}^i \in \check{X}_i$, follows directly from the relations $\bar{x}^i - w^i \in M$ and $M \in \mathcal{M}$, fulfilled for any $i \in N$ (remind, that by Sign-assumption, for any non-zero $x \in M$ with $M \in \mathcal{M}$ we have that $x_k < 0$ for some k). As to the second one, we apply the same argumentation, as in the proof of Proposition 2., which shows that supposition $w^i \in \mathcal{P}_i(\bar{x}^i)$ for some $i \in T_{\mathcal{E}}$ implies that coalition $\{i\}$ can w-improve the allocation $\bar{x}$. Applying the contradiction obtained (remind, it was supposed that $\bar{x}$ belongs to $D^M_{\mathcal{E}}$) and completeness of the preference relations α_i, $i \in T_{\mathcal{E}}$, we get required: $\bar{x}^i \in \alpha_i(w^i)$ for any $i \in T_{\mathcal{E}}$.

According to the remarks, given above, the proof of our theorem is completed. Q.E.D.

Corollary 3. *If $\mathcal{E}$ is C-regular with $\{i\} \in \sigma$, $i \in T_{\mathcal{E}}$, $w^i \in X_i$ for any $i \in N \setminus T_{\mathcal{E}}$, and, besides, for any $i \in N$ there exists a coalition $R \in \sigma_i$ such that*

$$(R \setminus \{i\}) \cap T_{\mathcal{E}} \neq \emptyset,$$

then $D_*(\mathcal{E}) \subseteq W(\mathcal{E})$.

We omit the proof of Corollary 3 because it almost literally reproduces the proof of Corollary 2 (with $S_{\mathcal{E}}$ replaced by $T_{\mathcal{E}}$).

As usual, we say that α_i is locally monotonic, if for any $x^i \in \mathrm{Pr}_{X_i} X(N)$ there exists $\delta(x^i) > 0$ such that $x^i + z \in \mathcal{P}_i(x^i)$, whenever $z \in \mathbf{R}^l_+$ and $0 < \|z\| < \delta(x^i)$. Observe, that α_i is always locally monotonic under the following standard assumptions: α_i is strongly monotonic, and $X_i = \mathbf{R}^l_+$.

Taking account that local monotonicity guarantees equilibrium prices to be strictly positive, and summarizing the results, which has been proven above, we obtain the following core equivalence result.

Theorem 5. *Suppose a pure exchange economy $\mathcal{E}$ satisfies the assumptions of either one of Theorems 3, 4, or Corollary 3. If α_i is locally monotonic for at least one agent of the economy $\mathcal{E}$, then $D_*(\mathcal{E}) = W(\mathcal{E})$.*

4. Conclusion

Two remarks should be given in the conclusion. First, let us mention once more that w-blocking introduced in the paper is not that much stronger than blocking considered in [9]. Although we have that directly by definition it follows that coalition S can block a feasible contractual M-system $\mathcal{V}$ whenever $\mathcal{V}$ can be w-blocked by S (i.e., w-blocking implies blocking, which means in our terms that the former is stronger than the latter), there are several rather wide classes of pure exchange economies with totally contractual set $D(\mathcal{E})$ and weak totally contractual core $D_*(\mathcal{E})$ to be equal. In fact, due to the inclusion $W^{\mathcal{M}}(\mathcal{E}) \subseteq D(\mathcal{E})$[1] established in [9], one can easily see that any condition, which provides insertion $D_*(\mathcal{E}) \subseteq W(\mathcal{E})$, guarantees the coincidence of $D(\mathcal{E})$ and $D_*(\mathcal{E})$. In particular, due to Theorem 5 we have the following assertion.

[1] Here $W^{\mathcal{M}}(\mathcal{E})$ consists of those allocations from $W(\mathcal{E})$ that can be supported by a strictly positive prices.

Proposition 3. *Let $\mathcal{E}$ satisfies the assumptions of either one of Theorems 3, 4, or Corollary 3. If α_i is locally monotonic for at least one agent of the economy $\mathcal{E}$, then $D(\mathcal{E}) = D_*(\mathcal{E})$.*

Almost the same argumentation, based on Corollary 2 and Proposition 2, yields the "individual" version of Proposition 3 (the only alteration we need in the corresponding proof is the replacement of $W(\mathcal{E}), D(\mathcal{E})$, and $D_*(\mathcal{E})$ by $W^M(\mathcal{E}), D^M(\mathcal{E})$, and $D_*^M(\mathcal{E})$ with $W^M(\mathcal{E}) := W(\mathcal{E}) \bigcap X_{\mathcal{E}}^M$, $M \in \mathcal{M}$).

Proposition 4. *If $\mathcal{E}$ is C_M-regular exchange economy with $w^i \in X_i$ for any $i \in N \setminus S_{\mathcal{E}}^M$, one-element coalition $\{i\}$ belongs to σ for any $i \in N$, and, moreover, for any $i \in N$ there exists a coalition $R \in \sigma_i$ such that*

$$(R \setminus \{i\}) \cap S_{\mathcal{E}}^M \neq \emptyset,$$

then $D^M(\mathcal{E}) = D_*^M(\mathcal{E})$.

Proposition 5. *Let $\mathcal{E}$ be a C_M-regular pure exchange model with w^i belonging to X_i for any $i \in N \setminus S_{\mathcal{E}}^M$. If $S_{\mathcal{E}}^M$ is a strongly σ-divisible subset of N, and one-element coalition $\{i\}$ belongs to σ for any $i \in S_{\mathcal{E}}^M$, then $D^M(\mathcal{E}) = D_*^M(\mathcal{E})$.*

Second (and the final) remark concerns the importance of the strong σ-divisibility in Theorem 5. To demonstrate the main assumption in the core equivalence theorem is relevant, two examples of pure exchange economies having unblocked (in the sense of Definition 3) allocations with no supporting equilibrium prices are given. The most interesting seems to be the second example with no equilibrium allocations and nonempty weak totally contractual core, exhibiting that the weak totally contractual allocation may be chosen as a compromise solution in case the classical market mechanism doesn't work.

Example 1. [$W(\mathcal{E}_1) \neq \emptyset$, $D_*(\mathcal{E}_1) \setminus W(\mathcal{E}_1) \neq \emptyset$]

Let $\mathcal{E}_1$ be pure exchange economy defined by the following parameters:

$$N = \{1,2,3,4,5\}, \quad \sigma = \{\{1,2\},\{2,3\},\{3,4,5\}\};$$

$$X_i = \mathbf{R}_+^2,\ i \in N, \quad w^1 = (3,0),\ w^2 = (6,0),\ w^3 = (6,1),\ w^4 = (9,1),\ w^5 = (6,3);$$

$$u_i(x_1,x_2) = \begin{cases} x_1 + 4x_2 & , \quad i = 1,2, \\ 2x_1 + 3x_2 & , \quad i = 3,4,5. \end{cases}$$

Put

$$M = \{x \in \mathbf{R}^2 \mid x_1 + 3x_2 = 0\}$$

and consider the allocation $\bar{x} = (\bar{x}^i)_{i \in N} \in X^M_{\mathcal{E}_1}$, defined as follows:

$$\bar{x}^1 = (0,1),\ \bar{x}^2 = (0,2),\ \bar{x}^3 = (9,0),\ \bar{x}^4 = (12,0),\ \bar{x}^5 = (9,2).$$

Note, that directly from Definition 6 it follows that any strongly σ-divisible coalition contains at least two participants. Hence, due to $N_{\bar{x}} = \{5\}$ we get that in our case $N_{\bar{x}}$ is not a strongly σ-divisible subset of N. Further, by analyzing restrictions of utility functions u_i to the corresponding intervals X^M_i, one can easily check that for any $i \in N \setminus \{5\}$ the bundle $\bar{x}^i$ is a maximal element on X^M_i w.r.t. the individual preference relation, generated by u_i. Consequently, there is only one coalition $S = \{3,4,5\} \in \sigma$ that may be able to w-improve some contractual system from $C(\bar{x})$. Suppose, it really w-improves $\bar{x}$. Since the functions u_i, $i = 3,4,5$, are strictly increasing w.r.t x_1 on the corresponding sets X^M_i, the definition of w-blocking implies that there exists $x \in X^M_{\mathcal{E}}$ such that $x^3_1 \geq \bar{x}^3_1$, $x^4_1 \geq \bar{x}^4_1$ and $x^5_1 > \bar{x}^5_1$. Consequently, $\sum\limits_{i \in S} x^i_1 > 30$, which contradicts to the relations $\sum\limits_{i \in S} x^i_1 < \sum\limits_{i \in N} x^i_1 = \sum\limits_{i \in N} w^i_1 = 30$. Therefore, we get $\bar{x} \in D^M_*(\mathcal{E})$.

At the same time, one can easily check, that the set $W(\mathcal{E}_1)$ of competitive equilibria of the economy $\mathcal{E}_1$ consists of the only allocation $\widehat{x} = (\widehat{x}^i)_{i \in N} \in X(N)$, given by the data:

$$\widehat{x}^1 = (0,5/3),\ \widehat{x}^2 = (0,10/3),\ \widehat{x}^3 = (39/5,0),\ \widehat{x}^4 = (54/5,0),\ \widehat{x}^5 = (57/5,0),$$

with the (normed) supporting equilibrium price vector $\widehat{p}$, given by the equality: $\widehat{p} = (5/14, 9/14)$.

Hence, we have got the required: $W(\mathcal{E}_1) \neq \emptyset$ and $D_*(\mathcal{E}_1) \setminus W(\mathcal{E}_1) \neq \emptyset$.

Example 2. $[\,W(\mathcal{E}_2) = \emptyset,\ D_*(\mathcal{E}_2) \neq \emptyset\,]$

Consider one more pure exchange economy $\mathcal{E}_2$, given by the data:

$$N = \{1,2,3\}, \quad \sigma = \{\{1,2\},\{2,3\}\};$$

$$X_i = \mathbf{R}^2_+,\ i \in N; \quad w^1 = (0,1),\ w^2 = (6,1),\ w^3 = (6,3);$$

$$u_i(x_1,x_2) = \begin{cases} x_1 + 4x_2 & , \quad i = 1, \\ x_1 & , \quad i = 2,3. \end{cases}$$

Put, as in the previous example, $M = \{x \in \mathbf{R}^2 \mid x_1 + 3x_2 = 0\}$, and pick the balanced allocation $\bar{x} = (\bar{x}^i)_{i \in N} \in M_{\mathcal{E}_2}(N)$ of the economy $\mathcal{E}_2$, defined as follows:

$$\bar{x}^1 = (0,1),\ \bar{x}^2 = (9,0),\ \bar{x}^3 = (3,4).$$

Due to the same reason, as in the first example, $N_{\bar{x}} = \{3\}$ is not a strongly σ-divisible set in $\mathcal{E}_2$. Further, suppose, $\bar{x} \notin D_*^M(\mathcal{E}_2)$. Doing, like in the previous example, one can easily check that the bundles $\bar{x}^1$ and $\bar{x}^2$ are maximal elements in X_1^M and X_2^M, respectively (w.r.t. the individual preferences, defined by the corresponding utility functions $u_1$and u_2). Consequently, the only coalition $S \in \sigma$, which might be w-improving for some contractual system from $C(\bar{x})$, is $S = \{2,3\}$. Suppose, that $S = \{2,3\}$ really w-improves some $\mathcal{V} \in C(\bar{x})$. By definition of w- improvement and construction of $\bar{x}$ and utility functions u_2, u_3, it means that there exists $x = (x^i)_{i \in N} \in X_{\mathcal{E}_2}^M$ such that $x_1^1 \geq 0, x_1^2 \geq 9$, and $x_1^3 > 3$. But these inequalities apparently contradict to the relations

$$\sum_{i \in N} x_1^i = \sum_{i \in N} w_1^i = 12,$$

following from the obvious inclusion $X_{\mathcal{E}_2}^M \subseteq X(N)$.

So, our economy $\mathcal{E}_2$ possesses at least one weak totally contractual M-allocation. As to the competitive equilibrium, from the famous "irreducibility" criterion for the linear exchange economies, proposed by D. Gale (see [2]), it follows immediately that $W(\mathcal{E}_2) = \emptyset$ (the latter fact can be verified directly, as well, just by applying the definition of equilibrium and taking account that coalition $S = \{2,3\}$ is too "self-sufficient"). Hence, we get required: $W(\mathcal{E}_2) = \emptyset$ and $D_*(\mathcal{E}_2) \neq \emptyset$.

Acknowlegments

The author express his gratitude to prof. S. Weber for useful discussion on the subject.
This work was supported by the Netherlands Organization for Scientific Research in the framework of the Russian-Dutch programme for scientific cooperation (NWO-grant 047.017.017). The author would like to appreciate also financial assistance from Russian Fund of Basic Research (grant 07-06-00363) and Russian Leading Scientific Schools Fund (grant 4999.2006.6).

References

[1] K.J. Arrow, F.H. Hahn, *General Competitive Analysis*, Amsterdam-New York-Oxford-Tokyo: North-Holland, 1991.

[2] D. Gale, The linear excange model, *Journ. Math. Econ.* **3** (1976), 205–209.

[3] W. Hildenbrand, A.P. Kirman, *Equilibrium Analysis*. Amsterdam-New York-Oxford-Tokyo: North-Holland, 1988.

[4] V.L. Makarov, On a concept of contract in the abstract economy, *Optimizatsija* **24** (1980), 5–17 (in Russian).

[5] V.L. Makarov, Coordination Models of Economic Interests, Novosibirsk State University Press: Novosibirsk, 1981 (in Russian).

[6] V.L. Makarov, Economical equilibrium: Existence and extremal properties. In: *Problems of the Modern Mathematics* (Gamkrelidze, R. V., ed.) 19 (1982), 23–59, VINITI: Moscow (in Russian).

[7] V.A. Vasil'ev, Exchange Economies and Cooperative Games, Novosibirsk State University Press: Novosibirsk, 1984 (in Russian).

[8] V.A. Vasil'ev, Cores and generalized NM-solutions for some classes of cooperative games. In: *Russian Contributions to Game Theory and Equilibrium Theory* (Driessen, T. S. H, G. van der Laan, V. A. Vasil'ev and E. B. Yanovskaya, eds), 2006, 91–149, Springer Verlag: Berlin-Heidelberg.

[9] V.A. Vasil'ev, Contractual *M*-core and equilibrium allocations. *J. Math. Sci.* **133**, no. 4 (2006), 1402–1409.

In: Perspectives in Applied Mathematics
Editor: Jordan I. Campbell, pp. 75-90
ISBN 978-1-61122-796-3

Chapter 5

A VERSION OF THE ELFVING OPTIMAL STOPPING TIME PROBLEM WITH RANDOM HORIZON

Elbieta Z. Ferenstein and Anna Krasnosielska
Faculty of Mathematics and Information Science
Warsaw University of Technology
Warsaw, Poland

1. Introduction

Suppose that there are m ordered companies which want to employ the best secretary and, at the same time, each secretary wants to work in the best company. Therefore, we can assume that all secretaries go for an interview to the best company first. Each secretary is assigned a value corresponding to her knowledge. Each company is allowed to hire only one secretary at the time of her appearance on the basis of the past and current observations and the assumed companies' ordering. At each time only one candidate may be presented. If more than one company wants to employ the same secretary at the same time, then the best one employs her. For the company the values of the secretaries are decreases in time. After a random time M the companies stop the interviewing process. In our model the secretaries appear according to a Poisson process.

Let us formalize the problem. Suppose that there are $m \geq 1$ ordered companies which will be called players. Their indices $1, 2, ..., m$ correspond to their ordering, called priority, so that 1 refers to the player with the highest priority (the best company) and m to the player with the lowest one (the least interesting company). Suppose that to the i-th secretary is assigned a value Y_i, called offer, corresponding to her knowledge and experience. The applicant with the highest value Y_i is considered the best. Y_i are *iid* nonnegative random variables with common piecewise continuous distribution function F. The players (the companies) observe sequentially offers Y_n (the values of the secretaries) at the jump times τ_n, $n = 1,2,...$, of a homogeneous Poisson process such that $\tau_n < M$, where M is independent of the τ's and Y's. Each player is allowed to accept only one of the values Y_i at the time of its appearance on the basis of the past and current observations and the assumed players' priorities. The observed Y values are time-sensitive with non-increasing discount function r. More precisely, any player who has just decided to take up an offer (that is, to accept the value Y_n) at $\tau_n < M$, gets the reward $Y_n r(\tau_n)\mathbb{I}(M > \tau_n)$ if and only if he has not obtained any reward before and there is no player with higher priority who has also decided to accept the current value Y_n. As soon as a player gets the reward, he quits the game. The remaining players accept offers in the same manner, their priorities are as previously. As soon as $\tau_n \geq M$, rewards of the players who have not finished the game yet are 0.

In real life the problem of finding the best secretary is often more complicated than the classical full-information secretary problem. Our model yields a better description than models without random horizon. For instance in a city there are more than one company that want to employ a secretary. Moreover, if a company does not hire a secretary by a random time M, it will lose its chance of hiring one, and their "payoff" will surely be zero.

Problems similar to the one described have been treated by many authors. Our game is a generalization, to the case of a multi-person game with random horizon, of the optimal stopping time problem formulated and solved first by Elfving (1967), and later considered also by Siegmund (1967) and generalized by Ferenstein and Krasnosielska (2007). Stadje (1987, 1990) considered the problem of selling k commodities of the same kind in finite time, where potential customers arrive according to a Poisson process. The aim of the owner is to maximize the discounted expected gain.

Enns and Ferenstein (1990) investigated a two-person nonzero-sum stopping game with priorities. In that game the offers are observed at jump times of a Poisson process. The player who selects, in a sequential way with priority constraints, a random variable which is larger than the one selected by his opponent wins 1, otherwise 0. So the aim of each of the players is to maximize the winning probability. Saario and Sakaguchi (1992) considered a multi-stopping best choice game with offers observed also at jump times of a Poisson process and with priority constraints till time T. Players' final rewards are maxima of their selected offers. Kurano, Nakagami and Yasuda (2000) also considered a game variant of a stopping problem for jump processes. Porosiski (2005) considered a zero-sum game version of the continuous time full-information best choice problem with positive random horizon. He obtained an equilibrium strategy in the case of a Poisson stream of rewards and exponential horizon. General models of multi-person discrete-time games with priorities were analyzed in Enns and Ferenstein (1987), Sakaguchi (1989) and Ferenstein (1993, 2006). Dixon (1993) considered three stopping games with different reward structures, and with Poisson stream of offers modeled as *iid* random variables with uniform distribution. Sakaguchi (2002) also considered a game of similar type. A two-person best choice full information game with imperfect observations was considered in Porosiski and Szajowski (1996).

The secretary problem is a well-known stopping-time problem discussed by many authors (see Ferguson (1989), Samuels (1991) and Nowak and Szajowski (1999) for numerous references). For example, Cowan and Zabczyk (1978) considered a continuous time a generalization of the secretary problem (see also Szajowski [25]). They investigated a random number of objects (apartments) in a finite time interval.

Sakaguchi (1986) studied the full information secretary problem for a Poisson process of arrivals on a time interval of random length. This means that in his model the decision-making process is restricted by a given random variable. Samuel-Cahn (1996) considered an optimal stopping time problem with independent random horizon. She considered a secretary problem with a known number of applicants for the job, such that after time M, which is a random "freeze time" variable with known distribution, it is impossible to hire an applicant. Samuel-Cahn showed that the optimal stopping problem with discrete-time and independent random horizon is essentially equivalent to a "discounted" fixed horizon optimal stopping problem derived from it.

The paper is organized as follows: In Section 2, we formulate a model of one-person game and prove a general theorem showing that an optimal stopping time problem with continuous time, discount function and random independent horizon is equivalent to an optimal stopping problem with a new discount factor. This theorem is a generalization to continuous time of the theorem presented by Samuel-Cahn (1996). We obtain the solution of our model and discuss its properties. Moreover, the influence of stochastic order and hazard rate order on the optimal mean reward is considered. In Section 3, a generalization of our one person game with random horizon to a multi-person game with random horizon is analyzed. Examples are presented in Section 4.

2. One-Person Game

In this section we generalize the Elfving stopping time problem to the case of a random horizon stopping time problem. The basic assumptions and notations follow those of Chow et al. (1971). Let $Y_1, Y_2, \ldots$ be independent, nonnegative random variables with common piecewise continuous distribution function F and mean $E(Y_1) \in (0, \infty)$. Let $0 < \tau_1 < \tau_2 < \ldots$ be the jump times of a homogeneous Poisson process $N(t), t \geq 0$, with intensity $\lambda > 0$. The random variables $Y_1, Y_2, \ldots$ are interpreted as offers observed successively by one player at times $\tau_1, \tau_2, \ldots$. We assume that the sequences $\{Y_n\}$, $\{\tau_n\}$ are independent. Moreover, assume that the horizon M is a positive random variable independent of the sequences $\{Y_n\}$, $\{\tau_n\}$ with known distribution function. Furthermore, a discount function $r : [0, \infty) \to [0, 1]$ is given, which is non-increasing, continuous from the right and $r(0) = 1$, $r(U) = 0$ and $r(u) \neq 0$ for $u \in [0, U)$, where U is finite or infinite. The reward for the player accepting Y_n at τ_n on the basis of the observations $Y_1, \ldots, Y_n, \tau_1, \ldots, \tau_n, \mathbb{I}(M > \tau_1), \ldots, \mathbb{I}(M > \tau_n)$ is $X_n = Y_n r(\tau_n)\mathbb{I}(M > \tau_n)$, where $\mathbb{I}(A)$ is the indicator function of the event A and M is positive random variable independent of the sequences Y's and τ's. Assume that $\int_0^\infty r(u)du < \infty$. Hence

$$E \sum_{n=1}^{\infty} Y_n r(\tau_n)\mathbb{I}(M > \tau_n) \leq E(Y_1) \int_0^\infty r(u)du < \infty \tag{1}$$

and as a consequence $E(\sup X_n) < \infty$ and $X_\infty = 0$.

Remark 1. *Taking a horizon M which satisfies $P(M = U) = 1$ we recover the original Elfving problem (see Chow [1] or Elfving [4]).*

Now we will formulate and prove the general theorem that an optimal stopping time problem with discount function and independent random horizon is equivalent to an optimal stopping problem with a modified discount function. Let us introduce various σ-fields of events dependent on available observations: $\mathcal{F}_i = \sigma(Y_1, \ldots, Y_i, \tau_1, \ldots, \tau_i)$; $\mathcal{F}_i^* = \sigma(Y_1, \ldots, Y_i, \tau_1, \ldots, \tau_i, \mathbb{I}(M >$

$\tau_1),...,\mathbb{I}(M>\tau_{i-1}))$; $\mathcal{F}_i^{**}=\sigma(Y_1,...,Y_i,\tau_1,...,\tau_i,\mathbb{I}(M>\tau_1),...,\mathbb{I}(M>\tau_i))$, $i=1,2,...$. Let $\mathcal{T}$, $\mathcal{T}^*$, $\mathcal{T}^{**}$ be the sets of stopping rules adapted to $\{\mathcal{F}_i\}$, $\{\mathcal{F}_i^*\}$, $\{\mathcal{F}_i^{**}\}$, respectively. Set $\mathbb{I}(M>\tau_i)=\xi_i$. The following considerations will be necessary to prove Theorem 1.. Note that for all $t^{**}\in\mathcal{T}^{**}$ there exist sequences $\{A_k\}_1^i$, $\{B_k\}_1^i$, $\{C_k\}_1^i$ such that $A_k\in\mathcal{B}(\mathcal{R}^k)$, $B_k\in\mathcal{B}(\mathcal{R}^k)$, $C_k\in\mathcal{B}(\mathcal{R}^k)$, $k=1,2,...,i$, and

$$\{t^{**}=i\}=\{\{Y_1\notin A_1\}\cup\{\tau_1\notin B_1\}\cup\{\xi_1\notin C_1\}\}\cap...$$

$$\cap\{\{(Y_1,...,Y_{i-1})\notin A_{i-1}\}\cup\{(\tau_1,..,\tau_{i-1})\notin B_{i-1}\}\cup\{(\xi_1,..,\xi_{i-1})\notin C_{i-1}\}\}\cap$$

$$\cap\{\{(Y_1,...,Y_i)\in A_i\}\cap\{(\tau_1,...,\tau_i)\in B_i\}\cap\{(\xi_1,..,\xi_i)\in C_i\}\},$$

where C_k is the set of all k-element sequences $(c_1,c_2,...,c_k)$ such that $c_1=...=c_{j-1}=1$, $c_j=...=c_k=0$, $j=2,...,k$ or $c_1=...=c_k=0$. Hence either $t^{**}=\inf\{i:\{(Y_1,...,Y_i)\in A_i\}\cap\{(\tau_1,...,\tau_i)\in B_i\}\cap\{\xi_i=1\}\}$ or $t^{**}=\inf\{i:\{(Y_1,...,Y_i)\in A_i\}\cap\{(\tau_1,...,\tau_i)\in B_i\}\cap\{\xi_i=0\}\}$. For given t^{**} define $t=\inf\{i:\{(Y_1,...,Y_i)\in A_i\}\cap\{(\tau_1,...,\tau_i)\in B_i\}\}$. Note that

$$\{t^{**}=i,\xi_i=1\}=\{t=i,\xi_i=1\}. \tag{2}$$

Theorem 1. *Let $Y_1,Y_2,...$ be iid nonnegative random variables and $X_n=g(Y_nr(\tau_n))\,\mathbb{I}(\tau_n<M)$, $E|X_n|<\infty$ for $n=1,2,...$. Let M be a positive random variable independent of the sequences of Y's and τ's. Moreover, assume that the Borel function $g:\mathcal{R}^+\cup 0\to\mathcal{R}^+\cup 0$, $g(0)=0$, g is nondecreasing and concave. Then, for any $k\in\mathcal{N}\cup\{0\}$,*

$$\sup_{t^{**}\in\mathcal{T}^{**}} E\Big(g(Y_{t^{**}}r(\tau_{t^{**}}))\mathbb{I}(M>\tau_{t^{**}})\mathbb{I}(t^{**}\geq k)\Big)= \tag{3}$$

$$=\sup_{t^*\in\mathcal{T}^*} E\Big(g(Y_{t^*}r(\tau_{t^*}))\mathbb{I}(M>\tau_{t^*})\mathbb{I}(t^*\geq k)\Big)=\sup_{t\in\mathcal{T}} E\Big(g(Y_tr(\tau_t))P(M>\tau_t)\mathbb{I}(t\geq k)\Big).$$

Proof. Clearly, $\mathcal{F}_i\subset\mathcal{F}_i^*\subset\mathcal{F}_i^{**}$, thus $\mathcal{T}\subset\mathcal{T}^*\subset\mathcal{T}^{**}$. Hence

$$\sup_{t\in\mathcal{T}} E\Big(g(Y_tr(\tau_t))P(M>\tau_t)\mathbb{I}(t\geq k)\Big)$$

$$\leq\sup_{t^*\in\mathcal{T}^*} E\Big(g(Y_{t^*}r(\tau_{t^*}))\mathbb{I}(M>\tau_{t^*})\mathbb{I}(t^*\geq k)\Big) \tag{4}$$

$$\leq \sup_{t^{**} \in \mathcal{T}^{**}} E\Big(g(Y_{t^{**}} r(\tau_{t^{**}}))\mathbb{I}(M > \tau_{t^{**}})\mathbb{I}(t^{**} \geq k)\Big).$$

Now, we show that for each $t^{**} \in \mathcal{T}^{**}$ one can find $t \in \mathcal{T}$ such that t and t^{**} yield the same reward. For any $t^{**} \in \mathcal{T}^{**}$, we have

$$g(Y_{t^{**}} r(\tau_{t^{**}}))\mathbb{I}(M > \tau_{t^{**}})\mathbb{I}(t^{**} \geq k) = \sum_{i=k}^{\infty} g(Y_i r(\tau_i))\mathbb{I}(M > \tau_i)\mathbb{I}(t^{**} = i)$$

$$= \sum_{i=k}^{\infty} g(Y_i r(\tau_i))\mathbb{I}(\xi_i = 1)\mathbb{I}(t^{**} = i) = \sum_{i=k}^{\infty} g(Y_i r(\tau_i))\mathbb{I}(\xi_i = 1)\mathbb{I}(t = i), \tag{5}$$

where the last equality is a consequence of (2). Thus we have shown that t is equivalent to t^{**} in the sense that they yield the same reward. Taking expectations on both sides of (5) and using properties of conditional expectation and the fact that M is independent of $\{Y_n\}$ and $\{\tau_n\}$ we obtain

$$E\Big(g(Y_{t^{**}} r(\tau_{t^{**}}))\mathbb{I}(M > \tau_{t^{**}})\mathbb{I}(t^{**} \geq k)\Big)$$

$$= E\Big(\sum_{i=k}^{\infty} E\Big(g(Y_i r(\tau_i))\mathbb{I}(M > \tau_i)\mathbb{I}(t = i)|Y_1, Y_2, \ldots, \tau_1, \tau_2, \ldots\Big)\Big)$$

$$= E\Big(g(Y_t r(\tau_t))\mathbb{I}(t \geq k)E\left(\mathbb{I}(M > \tau_t)|\tau_t\right)\Big) = E\Big(g(Y_t r(\tau_t))P(M > \tau_t)\mathbb{I}(t \geq k)\Big).$$

In the above equalities we have used (1) and Jensen's inequality. Now taking the supremum over all $t \in \mathcal{T}$ (or over all $t^{**} \in \mathcal{T}^{**}$), and using fact that t and t^{**} are equivalent in the sense that they yield the same reward, we get

$$\sup_{t^{**} \in \mathcal{T}^{**}} E\Big(g(Y_{t^{**}} r(\tau_{t^{**}}))\mathbb{I}(M > \tau_{t^{**}})\mathbb{I}(t^{**} \geq k)\Big) = \sup_{t \in \mathcal{T}} E\Big(g(Y_t r(\tau_t))P(M > \tau_t)\mathbb{I}(t \geq k)\Big).$$

Hence from (4) we have (3). ■

Remark 2. *The above theorem is true for a positive random variable M with any given distribution. Theorem 1 for the full-information classical secretary problem with discrete and curtailing random variable M, $g(x) = x$ and $k = 0$ was proved in Samuel-Cahn (1996).*

Corollary 1. *Suppose that the assumptions of Theorem 1 are satisfied and let* $u \geq 0$. *Then*

$$\sup_{t^{**} \in \mathcal{T}^{**}} E(Y_{t^{**}} r(u+\tau_{t^{**}}) \mathbb{I}(M > u+\tau_{t^{**}})) = \sup_{t \in \mathcal{T}} E(Y_t \tilde{r}(u+\tau_t)), \tag{6}$$

where

$$\tilde{r}(s) = r(s) P(M > s) \quad \text{for} \quad s \geq 0. \tag{7}$$

Proof. We apply Theorem 1 with $g(x) = x$ and $k = 0$. ∎

Remark 3. *Our optimal stopping problem with discount function and independent random horizon is equivalent to the optimal stopping problem with a new discount function defined in (7) in the sense that the optimal mean rewards are the same. The new discount function has the following properties:* $\tilde{r} : [0, \infty) \to [0, 1]$ *and* $\tilde{r}$ *is non-increasing, continuous from the right and* $\tilde{r}(0) = 1$, $\tilde{r}(\tilde{U}) = 0$, $\tilde{r}(u) \neq 0$ *for* $u \in [0, \tilde{U})$, *where* $\tilde{U} = \inf\{s > 0 : \tilde{r}(s) = 0\}$.

Let us introduce the σ-fields $\tilde{\mathcal{F}}_i^* = \sigma(Y_1, ..., Y_i, \tau_1, ..., \tau_i, \mathbb{I}(M \geq \tau_1), ..., \mathbb{I}(M \geq \tau_{i-1}))$ and $\tilde{\mathcal{F}}_i^{**} = \sigma(Y_1, ..., Y_i, \tau_1, ..., \tau_i, \mathbb{I}(M \geq \tau_1), ..., \mathbb{I}(M \geq \tau_i))$, $i = 1, 2, ...$ Let $\tilde{\mathcal{T}}^*$ and $\tilde{\mathcal{T}}^{**}$ be the sets of all stopping rules adapted to the sequences $\{\tilde{\mathcal{F}}_i^*\}$ and $\{\tilde{\mathcal{F}}_i^{**}\}$, respectively.

Proposition 1. *Suppose that the assumptions of Theorem 1 are satisfied. Moreover, let M be a continuous random variable. Then*

$$\sup_{t \in \mathcal{T}} E(g(Y_t r(\tau_t)) P(M > \tau_t)) = \sup_{\tilde{t}^* \in \tilde{\mathcal{T}}^*} E(g(Y_{\tilde{t}^*} r(\tau_{\tilde{t}^*})) \mathbb{I}(M \geq \tau_{\tilde{t}^*}))$$

$$= \sup_{\tilde{t}^{**} \in \tilde{\mathcal{T}}^{**}} E(g(Y_{\tilde{t}^{**}} r(\tau_{\tilde{t}^{**}})) \mathbb{I}(M \geq \tau_{\tilde{t}^{**}})).$$

Proof. Proof follows the lines of the proof of Theorem 1 and the fact that $P(M > \tau_t) = P(M \geq \tau_t)$. ∎

Note that putting $Z_n = (Y_n, \tau_n)$, we are in a homogeneous Markov case. Set $\tilde{U} = \inf\{s \geq 0 : \tilde{r}(s) = 0\}$, where $\tilde{r}(s)$ is given by (7). It is easy to see that $\tilde{U} \leq U$. Let $V(u)$ be the optimal mean reward for stopping the sequence $\{Y_n \tilde{r}(u + \tau_n)\}$ for $u \in (0, \tilde{U})$ with respect to the filtration $\mathcal{F}$. Let $V^{**}(u)$ be the optimal mean reward for stopping the sequence $\{Y_n r(u+\tau_n)\mathbb{I}(M > u+\tau_n)\}$ for $u \in (0,U)$ with respect to the filtration $\mathcal{F}^{**}$. Define the stopping times

$$\sigma_1 = \inf\{n \geq 1 : Y_n \tilde{r}(\tau_n) \geq V(\tau_n)\},$$

$$\sigma_1(u) = \inf\{n \geq 1 : Y_n \tilde{r}(u+\tau_n) \geq V(u+\tau_n)\},$$
$$\sigma_1^{**} = \inf\{n \geq 1 : Y_n r(\tau_n)\mathbb{I}(M > \tau_n) \geq V^{**}(\tau_n)\},$$
$$\sigma_1^{**}(u) = \inf\{n \geq 1 : Y_n r(u+\tau_n)\mathbb{I}(M > u+\tau_n) \geq V^{**}(u+\tau_n)\}.$$

Note that $\sigma_1^{**} = \sigma_1^{**}(0)$ is the optimal stopping time for the reward sequence $\{Y_n r(\tau_n)\mathbb{I}(M > \tau_n)\}$ and $\sigma_1 = \sigma_1(0)$ is the optimal stopping time for the reward sequence $\{Y_n \tilde{r}(\tau_n)\}$. Hence from Corollary 1 we obtain

$$V^{**}(u) = E(Y_{\sigma_1^{**}(u)} r(u+\tau_{\sigma_1^{**}(u)})\mathbb{I}(M > u+\tau_{\sigma_1^{**}(u)})) =$$

$$= \sup_{t^{**} \in \mathcal{T}^{**}} E(Y_{t^{**}} r(u+\tau_{t^{**}})\mathbb{I}(M > u+\tau_{t^{**}})) = \sup_{t \in \mathcal{T}} E(Y_t \tilde{r}(u+\tau_t)) =$$

$$= E(Y_{\sigma_1(u)} \tilde{r}(u+\tau_{\sigma_1(u)})) = V(u).$$

Set

$$y(u) = \frac{V(u)}{\tilde{r}(u)} \qquad \text{if} \qquad 0 \leq u < \tilde{U}. \tag{8}$$

If $u \geq \tilde{U}$, then $y(u) = 0$. Define $G(y) = 1 - F(y)$, $H(y) = \int_y^\infty y' dF(y')$ and $f_{1,u}(v) = P(\tau_{\sigma_1(u)} > v)$. Elfving obtained the formula (see Chow (1971 pp. 114-115))

$$f_{1,u}(v) = \exp\left[-\int_u^{u+v} \lambda G(y(v'))dv'\right], \quad \text{for} \quad 0 \leq v \leq \tilde{U} - u. \tag{9}$$

Hence, one derives the optimal mean reward

$$y(u)\tilde{r}(u) = V(u) = \lambda \int_u^{\tilde{U}} \tilde{r}(v) H(y(v)) \exp\left[-\int_u^v \lambda G(y(v'))dv'\right] dv. \tag{10}$$

Differentiating both sides of (10) with respect to u we obtain

$$\frac{d}{du}[\tilde{r}(u)y(u)] = \lambda \cdot \tilde{r}(u) \cdot \Big(y(u)G(y(u)) - H(y(u))\Big). \tag{11}$$

The last equation is a first order non-linear differential equation which can be solved numerically with the condition $\tilde{r}(u)y(u) \to 0$ as $u \to \tilde{U}$ (see Chow (1971, pp. 115-117)). Note that $y(0) = \lim_{u \to 0^+} y(u)$ because $y(\cdot)$ is right continuous since $\tilde{r}(\cdot)$ is right continuous and $V(\cdot)$ is continuous. The theorem below tells us how to find the optimal mean reward and optimal stopping time.

Theorem 2. *(i) A piecewise continuous function $\tilde{y}(\cdot)$ satisfies (10) if and only if $\tilde{y}(\cdot)$ satisfies (11) and $\tilde{r}(u)\tilde{y}(u) \to 0$ as $u \to \tilde{U}$.*
(ii) If $\tilde{y}(\cdot)$ satisfies (10) on $(0, \tilde{U})$ and $\tilde{y}(\cdot) = 0$ on $[\tilde{U}, \infty)$, then $\tilde{y}(\cdot) = y(\cdot)$ in (8), that is, it determines $V(\cdot)$ and $\sigma_1 = \inf\{n \geq 1 : Y_n \geq \tilde{y}(\tau_n)\}$.

Remark 4. *$\sigma_1(0)$ is the optimal stopping rule, $\tau_{\sigma_1(0)}$ is the optimal stopping time and $V(0)$ is the optimal mean reward for the player.*

Remark 5. *Without loss of generality we can consider the game with a homogeneous Poisson process because a non-homogeneous Poisson process with intensity function $p(\cdot)$ can be reduced to a homogeneous Poisson process (see Chow (1971, pp. 113-114)).*

Below we analyze the influence of the horizon's order on the optimal mean rewards. Let M_1 and M_2 be positive random variables independent of the sequences $\{Y_n\}$, $\{\tau_n\}$ with known distribution functions F_1 and F_2, respectively. Let us recall the following definitions.

Definition 1. *(stochastic order) We say that M_1 is smaller than M_2 in the stochastic order (denoted by $M_1 \prec_{st} M_2$) if $F_1(x) \geq F_2(x)$ for all real x.*

Definition 2. *(hazard rate order) We say that M_2 is smaller than M_1 in the hazard rate order (denoted $M_2 \prec_{hr} M_1$) if the function $x \mapsto (1 - F_1(x))/(1 - F_2(x))$ is increasing.*

Let $V^{M_1}(u)$ and $V^{M_2}(u)$ be the optimal mean rewards for the player in the above one-person games with random horizons M_1 and M_2, respectively.

Theorem 3. *(i) If $M_1 \prec_{st} M_2$, then $V^{M_1}(0) \leq V^{M_2}(0)$.*
(ii) Assume additionally that M_1 and M_2 have continuous distribution functions. Let $U = \min\{m_1, m_2\}$ and $m_i = \inf\{s : F_i(s) = 1\}$, $i = 1, 2$. If $M_2 \prec_{hr} M_1$, then $V^{M_2}(0) \leq V^{M_1}(0)$.

Proof. (i) From the assumption we have $F_1(m) \geq F_2(m)$, hence $P(M_1 > m) \leq P(M_2 > m)$. Therefore $Y_t\tilde{r}^{M_1}(\tau_t) \leq Y_t\tilde{r}^{M_2}(\tau_t)$, where $\tilde{r}^{M_i}(\tau_t) = r(\tau_t)P(M_i > \tau_t)$, $i = 1,2$. To complete the proof of (i) it is sufficient to take expectation on both sides of the last inequality and then the supremum over all $t \in \mathcal{T}$.
(ii) Under assumptions of the theorem, the hazard rate order implies the stochastic order (see Müller and Stoyan (2002, p. 11)). ■

Remark 6. *Theorem 3 (i) shows that the optimal mean reward in the original Elfving problem is greater than or equal to the optimal mean reward in the Elfving problem with the same parameters and with an additional random horizon M.*

Now we will analyze the influence of different values of λ on rewards. Let $V^{\lambda_1}(u)$ and $V^{\lambda_2}(u)$ be the optimal mean rewards in our one-person game with intensity λ_1 and λ_2, respectively.

Proposition 2. *Suppose Y_1 has continuous distribution function, $\tilde{r}$ is differentiable on $(0,\tilde{U})$ and $\tilde{r}(u) \neq 0$ for $u \in (0,\tilde{U})$, where $\tilde{U} = \inf\{s : \tilde{r}(s) = 0\}$. Then $V^{\lambda_1}(0) \geq V^{\lambda_2}(0)$ for $\lambda_1 > \lambda_2$.*

Proof. Let us denote $y^{\lambda_i}(u) = V^{\lambda_i}(u)/r(u)$, $i = 1,2,$, for $u \in (0,U)$ and 0 otherwise. Let us rewrite the equation (11) for $i = 1,2$ in the following form

$$\frac{d}{du}y^{\lambda_i}(u) = \lambda_i \cdot \left(y^{\lambda_i}(u)G(y^{\lambda_i}(u)) - H(y^{\lambda_i}(u))\right) - \frac{\frac{d}{du}r(u)}{r(u)}y^{\lambda_i}(u). \tag{12}$$

Taking $u = -t + U$ we get differential equation

$$\frac{d}{dt}y^{\lambda_i}(-t+U) = -\frac{\frac{d}{dt}r(-t+U)}{r(-t+U)}y^{\lambda_i}(-t+U) \tag{13}$$

$$+\lambda_i \cdot \left(H(y^{\lambda_i}(-t+U)) - y^{\lambda_i}(-t+U)G(y^{\lambda_i}(-t+U))\right).$$

with the boundary condition $y^{\lambda_i}(U) = 0$. Let us note that $H(y^{\lambda_i}(-t+U)) - y^{\lambda_i}(-t+U)G(y^{\lambda_i}(-t+U)) \geq 0$. Hence, using Gronwall Lemma we get $y^{\lambda_1}(-t+U) \geq y^{\lambda_2}(-t+U)$, which can be rewritten as $y^{\lambda_1}(u) \geq y^{\lambda_2}(u)$. Therefore $V^{\lambda_1}(0) = V^{\lambda_1}(0)$. ■

3. Multi-person Game

Now we will generalize the multi-person game considered in Ferenstein and Krasnosielska (2007) to the case of a multi-person game with random horizon. There are $m > 1$ ordered players who sequentially observe offers Y_n at times τ_n, $n = 1, 2, \ldots$. Players' selection strategies based on their priorities are described in Introduction. The players' rewards are $Y_n r(\tau_n)\mathbb{I}(M > \tau_n)$ and observable events are from $\{\mathcal{F}_n^{**}\}$. As in the one-person game the above game is equivalent to the game with rewards $\{Y_n \tilde{r}(\tau_n)\}$ with respect to $\{\mathcal{F}_n\}$.

Let $V_{m,k}(u)$ be the optimal mean reward for the game with reward sequence $\{Y_n \tilde{r}(u + \tau_n)\}$ for Player k in the m-person game, which starts at time u, $u \geq 0$, where k is the player's priority, $k = 1, \ldots, m$, $m = 2, 3, \ldots$. Note that $V_{1,1}(u) = V(u)$ and $y_{1,1}(u) = y(u)$. The following theorems (see Ferenstein and Krasnosielska (2007)) allow one to compute the optimal mean rewards of the players in the m-person game for the sequence $\{Y_n \tilde{r}(u + \tau_n)\}$ with respect to $\{\mathcal{F}_n\}$.

Theorem 4. *The optimal mean reward for Player $m > 1$ in the m-person game satisfies the equation*

$$\frac{d}{du} V_{m,m}(u) = \frac{d}{du}[\tilde{r}(u) y_{m,m}(u)]$$

$$= \lambda \cdot \Big[\tilde{r}(u) \cdot \Big(G(y_{m,m}(u)) y_{m,m}(u) - H(y_{m,m}(u)) + H(y_{m-1,m-1}(u))\Big)$$

$$- V_{m-1,m-1}(u) G(y_{m-1,m-1}(u))\Big], \tag{14}$$

where $0 \leq u < \tilde{U}$ and $y_{m,m} = V_{m,m}(u)/\tilde{r}(u)$.

Equation (14) is a first order non-linear differential equation which can be solved numerically with the condition $\tilde{r}(u) y_{m,m}(u) \to 0$ as $u \to \tilde{U}$.

Theorem 5. *The optimal mean reward for Player k, $k = 1, 2, \ldots, m-1$, in the m-person game starting at $u \geq 0$, $m = 2, 3, \ldots$, is equal to the optimal mean reward of Player k in the k-person game: $V_{m,k}(u) = V_{k,k}(u)$ for $k = 1, 2, \ldots, m-1$ and $u \geq 0$.*

The statement of the theorem above is intuitively clear. The decision of Player k, $k = 1, 2, \ldots, m-1$, is not influenced by the decisions of players with lower priorities.

Remark 7. *From Theorem 5 we can get by recursion the optimal mean reward for Player k, $k = 1,...,m-1$, in the m-person game.*

Remark 8. *The optimal mean reward for Player k in the m-person game is $V_{m,k}(0)$, $k = 1,...,m-1$, $m = 1,2,....$*

4. Examples

Let $\{Y_i\}$ be the sequence of *iid* r.v's with exponential distribution with mean 1. Assume that $r(u) = 1$, $u \in (0,U)$, $U = 3$ and $\lambda = 1$. Then $F(y(u)) = 1 - \exp(-y(u))$, $G(y(u)) = \exp(-y(u))$, $H(y(u)) = (y(u)+1) \cdot \exp(-y(u))$.

Example 1. *Assume that $P(M = U) = 1$ (the original Elfving problem). Hence $\tilde{r}(u) = 1$ for $u \in (0,U)$ and $U = \tilde{U}$. Therefore the differential equation (11) becomes*

$$\frac{dy(u)}{du} = -\exp(-y(u)). \tag{15}$$

Solving the equation (15) with the boundary condition $y(U) = 0$ we get $V(u) = y(u) = \ln(1+U-u)$. Hence for $U = 3$, the optimal mean reward is $V(0) = \lim_{u \to 0} V(u) = \ln(1+U) = \ln 4$.

Example 2. *Assume that M has Weibull distribution with parameters $c > 0$ and $\alpha > 0$ (Weibull(c,α)). Then $P(M > u) = \exp(-cu^{\alpha})$. Hence $\tilde{r}(u) = r(u) \cdot P(M > u) = 1 \cdot \exp(-cu^{\alpha}) = \exp(-cu^{\alpha})$, where $u \in (0,\tilde{U})$. The differential equation (11) becomes*

$$-\alpha \cdot c \cdot u^{\alpha-1} \cdot y(u) + y'(u) = -\exp(-y(u)). \tag{16}$$

Example 3. *Assume that M has gamma distribution with parameters $(2,\beta)$, $\beta > 0$, (gamma$(2,\beta)$). Then $P(M > u) = (\beta u+1) \cdot \exp(-\beta u)$. Therefore $\tilde{r}(u) = r(u) \cdot P(M > u) = 1 \cdot (\beta u+1) \cdot \exp(-\beta u) = (\beta u+1) \cdot \exp(-\beta u)$, where $u \in (0,\tilde{U})$. The differential equation (11) becomes*

$$-\beta^2 u \cdot y(u) + (\beta u+1) \cdot y'(u) = -(\beta u+1) \cdot \exp(-y(u)). \tag{17}$$

Using numerical methods with the boundary condition $y(U) = 0$ we can solve equations (16) and (17).

Example 4. *Let M have the following distribution: $P(M = 1/2) = 1/3$ and $P(M=2) = 2/3$. Hence $\tilde{U} = 2$. Therefore we have the differential equations*

$$\frac{dy(u)}{du} = -\exp(-y(u)) \quad for \quad u \in \left(0, \frac{1}{2}\right), \tag{18}$$

$$\frac{2}{3} \cdot \frac{dy(u)}{du} = -\frac{2}{3}\exp(-y(u)) \quad for \quad u \in \left(\frac{1}{2}, 2\right). \tag{19}$$

Solving (19) with the boundary condition $y(\tilde{U}) = 0$ we get $y(u) = \ln(2-u)$ for $u \in (1/2, 2)$. Note that the solution of (18) is $y(u) = \ln(C-u)$ for $u \in (0, 1/2)$, where $C \in \mathcal{R}$. From (10) we see that $V(u) = \tilde{r}(u)y(u)$ is continuous for all $u \in [0, \tilde{U})$, where

$$\tilde{r}(u) = \begin{cases} 1 & for \quad u \in \left(0, \frac{1}{2}\right), \\ \frac{2}{3} & for \quad u \in \left(\frac{1}{2}, 2\right), \\ 0 & otherwise. \end{cases}$$

Hence $\lim_{u \to (1/2)^-} \tilde{r}(u)\ln(C-u) = \lim_{u \to (1/2)^+} \tilde{r}(u)\ln(2-u)$, *so* $C = (3/2)^{2/3} + 1$. *Therefore* $V(0) = \ln[(3/2)^{2/3} + 1]$.

In Tables 1 and 2 the optimal values $V(0)$ are given for various distributions of M and $U = 3$, $\lambda \in \{1, 5\}$.

Table 1. Optimal mean rewards for $V(0)$ for M with gamma and Weibull distributions.

U=3		$\lambda = 1$	$\lambda = 5$	U=3		$\lambda = 1$	$\lambda = 5$
gamma	E(M)	$V(0)$	$V(0)$	Weibull	E(M)	$V(0)$	$V(0)$
(10,1)	10	1.38617	2.77237	(3,0.1)	61.454	.075069	.159747
(3,1)	3	1.1428	2.33982	(1,0.3)	9.26053	.528631	1.10205
(2,1)	2	.917704	1.9598	(1,0.5)	2	.541009	1.17552
(2,2)	1	.616614	1.49031	(1,2)	.886227	.593492	1.49726
(5,5)	1	.657335	1.63172	(3,10)	.852371	.613851	1.64257
(3,5)	.6	.441824	1.22528	(5,2)	.396333	.320867	.99126
(10,20)	.5	.398182	1.19824	(5.22,1.5)	.3	.249047	.8027
(3,10)	.3	.252535	.830105	(3,0.3)	.237811	.107699	.282984
(2,9)	.222222	.192023	.661916	(3,0.5)	.222222	.151912	.438733
(5,50)	.1	.094431	.391664				

Table 2. Optimal mean rewards for $V(0)$ for M with exponential distributions.

U=3		$\lambda = 1$	$\lambda = 5$
Distribution	E(M)	$V(0)$	$V(0)$
Deterministic	3	1.38629	2.77259
Exponential(0.5)	2	.808976	1.72898
Exponential(1)	1	.562505	1.32587
Exponential(2)	.5	.351626	.958561
Exponential(4.5)	.222222	.18473	.606062

The graphs of the optimal mean rewards for the parameters $\lambda = 5$, $E(M) = 2$ and $E(M) = 2/9$ are displayed respectively in Figure 1.

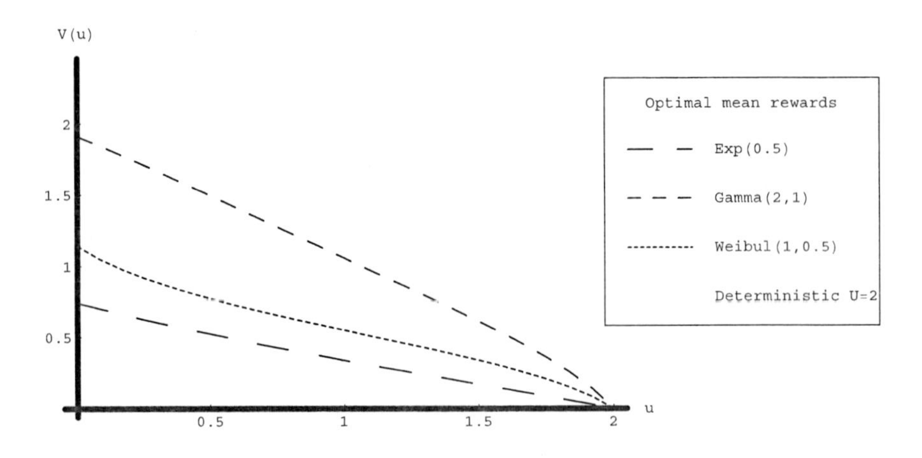

Figure 1. The optimal mean rewards for the player in one person game, where the random horizon has expectation $E(M) = 2$.

References

[1] Y.S. Chow, H. Robbins, D. Siegmund, *Great Expectations: The Theory of Optimal Stopping*, Boston: Houghton Mifflin, 1971.

[2] R. Cowan, J. Zabczyk, An Optimal Selection Problem Associated With The Poisson Process, *Theory of Probability and its Applications* **23** (1978), 584–592.

[3] M.T. Dixon, Equilibrium Points for Three Games Based on the Poisson Process, *Journal of Applied Probability* **30** (1993), 627–638.

[4] G. Elfving, A Persistency Problem Connected With a Point Process, *Journal of Applied Probability* **4** (1967), 77–89.

[5] E.G. Enns, E. Z. Ferenstein, On a Multi-Person Time-Sequential Game With Priorities, *Sequential Analysis* **6** (1987), 239–256.

[6] E.G. Enns, E. Z. Ferenstein, A Competitive Best-Choice Problem With Poisson Arrivals, *Journal of Applied Probability* **27** (1990), 333–342.

[7] E.Z. Ferenstein, A Variation of the Dynkin's Stopping Game, *Mathematica Japonica* **38** (1993), 371–379.

[8] E.Z. Ferenstein, On Infinite-Horizon Multi Stage Stopping Games With Priorities, in *Game Theory and Mathematical Economics*, eds. A. Wieczorek, M. Malawski and A. Wiszniewska-Matyszkiel, Warsaw: Banach Center Publications 71 (2006), 93–102.

[9] E.Z. Ferenstein, A. Krasnosielska *Nash Equilibrium in a Game Version of Elfving Problem*, in press.

[10] T. Ferguson, Who Solved The Secretary Problem?, *Statistical Science* **4** (1989), 282–296.

[11] M. Kurano, J. Nakagami, M. Yasuda, Game Variant of Stopping Problem on Jump Processes With a Monotone Rule, in *Advances in Dynamic Games and Applications*, eds. J. A. Filar, V. Gaitsgory and K. Mizumaki, Boston: Birkhäuser, 2000, 257–266.

[12] A. Müller, D. Stoyan, *Comparison Methods for Stochastic Models and Risks*, Chichester: John Wiley and Sons, 2002.

[13] A.S. Nowak, K. Szajowski, Nonzero-Sum Stochastic Games, *Annals of the International Society of Dynamic Games* 4 (1999), 297–343.

[14] Z. Porosiski, Modified Strategies in a Competitive Best Choice Problem With Random Priority, *Annals of the International Society of Dynamic Games* 7 (2005), 263–270.

[15] Z. Porosiski, On optimal choosing of one of the k best objects, *Statistic and Probability Letters* **65** (2003), 419–432.

[16] Z. Porosicski, K. Szajowski, On Continous-Time Two Person Full-Information Best Choice Problem With Imperfect Observation, *Sankhyā: The Indian Journal of Statistics Ser. A*, **58** (1996), 186–193.

[17] V. Saario, M. Sakaguchi, Multistop Best Choice Games Related to the Poisson Process, *Mathematica Japonica* **37** (1992), 41–51.

[18] M. Sakaguchi, Best Choice Problems for Randomly Arriving Offers During a Random Lifetime, *Mathematica Japonica* **31** (1986), 107–117.

[19] M. Sakaguchi, Multiperson Multilateral Secretary Problems, *Mathematica Japonica* **34** (1989), 459–473.

[20] M. Sakaguchi, Better-Than-Opponent: A Stopping Game for Poisson-Arriving Offers, *Scientia Mathematica Japonica* **56** (2002), 457–473.

[21] E. Samuel-Cahn, Optimal Stopping With Random Horizon With Application to the Full-Information Best-Choice Problem With Random Freeze, *Journal of American Statistical Association* **91** (1996), 357–364.

[22] S.M. Samuels, Secretary Problems, in *Handbook of Sequential Analysis*, eds. B. K. Gosh and P. K. Sen, New York: Marcel Dekker, 1991, 381–405.

[23] D. Siegmund, Some Problems in the Theory of Optimal Stopping, *The Annals of Mathematical Statistics* **38** (1967), 1627–1640.

[24] W. Stadje, An Optimal k-Stopping Problem for the Poisson Process, in *Proceedings of the 6th Pannonian Symposium on Mathematical Statistics*, B, eds. P. Bauer, F. Konecny and W. Wertz, Dordrecht: Reidel, 1987, 231–244.

[25] W. Stadje, A Full Information Pricing Problem for the Sale of Several Identical Commodities, *Mathematical Methods of Operations Research* **34** (1990), 161–181.

[26] K. Szajowski, *A Game Version of the Cowan-Zabczyk-Bruss' Problem*, in press.

In: Perspectives in Applied Mathematics ISBN 978-1-61122-796-3
Editor: Jordan I. Campbell, pp. 91-115

Chapter 6

HOMOTOPIES ON PREFERENCES UNDER ASYMMETRIC INFORMATION

Debora Di Caprio [*] ***and Francisco J. Santos-Arteaga*** [†]
School of Economics and Management,
Free University of Bozen-Bolzano,
Via Sernesi 1 - P.O. Box 276, 39100 Bolzano, Italy

Abstract

The present paper provides a strategic dynamic analysis of a theoretical scenario where *verifiable* information is transmitted between an informed sender and an uninformed but rational decision maker in a multidimensional space. The information provided by the sender is assumed to be encoded in *multifunctions*. We show that each one of these multifunctions induces a preference relation on the decision maker. These induced preferences do not generally coincide with those defined in a complete information environment. We provide sufficient conditions for the utility functions induced by the information sender to be continuous. In addition, we use homotopy theory to illustrate how the information encoded by the sender in the multifunctions can be modified through time in a continuous way so as to induce *any* a priori assigned preference relation on the decision maker.

2000 AMS: 91B06; 91B08; 91B44; 91A44; 55P10

---

[*]E-mail address: DDiCaprio@unibz.it, dicaper@mathstat.yorku.ca. Phone: +39 0471013168; Fax: +39 0471013009. (Corresponding author).

[†]E-mail address: FSantosArteaga@unibz.it, jarteaga@econ.yorku.ca

1. Introduction

Homotopy theory has been sporadically applied to economic theory mainly in order to simplify the aggregation of preferences among decision makers in social choice (see [17]), and to design stable algorithms in computable general equilibrium models (see [14]). These applications, while dealing with relevant issues, do not consider explicitly the influence of information asymmetries on the behaviour of decision makers, which constitutes a leading argument in current economic theory.

The strategic analysis derived from a scenario where *unverifiable* information is transmitted between an informed sender and an uninformed but rational decision maker was first introduced by Crawford and Sobel ([4]) in the economic literature. In their seminal model an unilaterally informed agent, who observes privately a signal realization regarding a variable defined by a unique characteristic, sends a message to a receiver who takes an action determining the wealth of them both. The main objective of their model - and the subsequent cheap talk literature - was the design of a mechanism that would induce the full revelation of all transmitted information. If such a mechanism cannot be correctly defined, all decisions taken by a given decision maker would be prone to manipulation by the information sender.

Sufficient conditions for the existence of a mechanism that induces the full revelation of all transmitted information when a unique unknown characteristic is considered are provided by Dulleck and Kerschbamer ([13]). Moreover, the multidimensional cheap talk literature has partially generalized the existence of full revelation mechanisms to settings where either multiple one-dimensional real variables or a single multi-dimensional real variable are considered, see [3] and [1], respectively. This branch of the literature concentrates efforts on the design of mechanisms that allow decision makers to elicit as much information as possible from senders.

The revelation mechanisms defined in these papers depend crucially on the similarity of the preference orders with which the sender and the decision maker are endowed. These preferences are exogenously given and cannot be manipulated by the information sender. In addition, the described literature assumes the common knowledge of all utility functions, preference orders, and probability density functions (i.e. subjective beliefs) on the set of all variables, or goods, before being able to derive a full revelation mechanism. The Bayesian Nash equilibrium on which the full revelation mechanism is based requires also a set of probability density functions to be defined on the information being transmitted. We do not consider such a requirement, since all transmitted information will be assumed verifiable by the decision maker.

An information sender with full knowledge regarding the preferences and beliefs of the decision maker should be able to manipulate them, leading to new more complex strategic scenarios when designing the corresponding revelation mechanisms. Indeed, this should be the case independently of the ability of the decision maker to verify the information transmitted by the sender. For example, Huffman et al. ([16]) analyze the modification of decision makers' beliefs and choices through the arrival of new information about genetically modified food. The main conclusion derived from their paper is that decision makers are indeed susceptible to information from interested third parties *even if such information is verifiable*.

The main purpose of the current paper is to define a new theoretical structure that allows for a dynamic analysis of preference manipulation in multidimensional spaces when all transmitted information is verifiable. In order to do so, we generalize the space of analysis, usually restricted to a finite-dimensional real vector space, to a generic product of abstract spaces. Besides, the use of homotopy techniques adds a dynamic dimension to the manipulation problem that allows the model to account for learning and limited memory phenomena.

The information provided by the sender, which is completely verifiable by the decision maker, is assumed to be encoded in *multifunctions* (i.e. set-valued maps). We show that each one of these multifunctions induces a preference relation on the decision maker. The preferences induced in this way will not generally coincide with those defined in a complete information environment, where all variables and their characteristics are known to decision makers.

We provide sufficient conditions for the utility functions induced by the information sender on the decision maker to be continuous. In this way, the main requirement[1] imposed by the cited literature to guarantee the existence of a full revelation mechanism would be satisfied but the mechanism remains manipulable. Indeed, we show that the common knowledge assumption regarding utilities and probability functions allows for the manipulation of the decision maker's preferences *before* the strategic information transmission process modelled in [4], [1], and [3] takes place. This is due to the fact that, even though the information sender is not allowed to lie, he is still able to display the information

[1]The other main condition, not necessary in our setting, is the compactness of the domain of the utility functions.

subsets he finds more convenient.[2]

Finally, we use homotopy theory to illustrate how the information encoded by the sender in the multifunctions can be modified through time so as to induce *any* a priori assigned preference relation on the decision maker. More precisely, we investigate the conditions for the existence of a homotopy such that the preferences of the decision maker can be directed in a smooth (i.e. continuous) way towards *any* predetermined good.

The paper proceeds as follows. Sections 2 and 3 introduce the notations and assumptions needed to develop the model. Sections 4 defines info-multifunctions, info-maps and their corresponding induced preferences. Sufficient conditions for these preferences to be continuously representable are discussed in Section 5. Section 6 develops the notions of compatible dilatations and shrinkings for an info-multifunction, leading to the concept of homotopic induced preferences. A particularly interesting class of compatible dilatations and shrinkings is studied in Section 7. Finally, Section 8 presents some concluding remarks.

2. Preliminaries and Basic Notations

Let X be a nonempty set. A **preference relation** on X is a binary relation $R \subseteq X \times X$ satisfying:

reflexivity: $\forall x \in X,\ (x,x) \in R$;

completeness: $\forall x,y \in X,\ (x,y) \in R \ \vee\ (y,x) \in R$;

transitivity: $\forall x,y,z \in X,\ (x,y) \in R \ \wedge\ (y,z) \in R \ \Rightarrow\ (x,z) \in R$.

Preference relations are usually denoted by the symbol $\succsim$. We will write $x \succsim y$ in place of $(x,y) \in \succsim$ and read: *x is preferred or indifferent to y*.

There two preference relations usually associated to a preference relation $\succsim$, the strict preference relation, defined by $x \succ y \overset{def}{\Longleftrightarrow} x \succsim y \ \wedge\ y \not\succsim x$, and the indifference relation, defined by $x \sim y \overset{def}{\Longleftrightarrow} x \succsim y \ \wedge\ y \succsim x$. We read $x \succ y$ as *x is preferred to y*, while $x \sim y$ is read *x is indifferent to y*.

[2]Economic decision makers are generally assumed to be female, and we will refer to them as such throughout the paper. On the other hand, information senders are assumed to be male.

From the definition it is clear that preference relations are complete preorders. Also, preference relations that are complete and transitive are usually called *rational*. Hence, all the preference relations in this paper are rational.

In particular, the symbols $\geq$ and $>$ will denote the standard partial and linear order on the reals, respectively.

A **utility function** representing $\succsim$ is any function $u : X \to \mathbb{R}$ such that:

$$\forall x, y \in X,\ x \succsim y \iff u(x) \geq u(y).$$

It is known that any (rational) preference relation $\succsim$ on a nonempty set X can be represented by a utility function if and only if it is *perfectly separable*, that is, if there exists a countable subset V of X such that for all $x \succ y$ there exists $z \in V$ with $x \succsim z \succsim y$ (see [25]).

If X is endowed with a topology τ, the problem of the existence of a continuous utility function representing $\succsim$ can be considered.[3]

Henceforth, we will let $\mathcal{G}$ denote **the set of all goods**, or commodities, and fix $n \geq 2$. Moreover, for every $i \leq n$, X_i will represent the set of all possible variants for the i-th characteristic or attribute of any commodity in $\mathcal{G}$, while X **will stand for the Cartesian product** $\prod_{i \leq n} X_i$.

Thus, an element $x_i^G \in X_i$ specifies the i-th characteristic of a given good $G \in \mathcal{G}$, while an n-tuple $(x_1^G, \ldots, x_n^G)$ lists all its characteristics.

For every $i \leq n$, X_i will be called the **i-th characteristic factor**. The Cartesian product $X = \prod_{i \leq n} X_i$ will be referred to as the **characteristic space**.

The preference relation on X will depend on the preference relations defined on the characteristic factors according to the assumptions presented in the following section.

A preference relation $\succsim$ on X is called **additive** (see [26]) if it is representable by an additive utility function, that is, if there exist $u : X \to \mathbb{R}$ and $u_i : X_i \to \mathbb{R}$, where $i \leq n$, such that $\forall (x_1, \ldots, x_n), (y_1, \ldots, y_n) \in X$,

$$u(x_1, \ldots, x_n) = u_1(x_1) + \cdots + u_n(x_n)$$

and

$$(x_1, \ldots, x_n) \succsim (y_1, \ldots, y_n) \Leftrightarrow u(x_1, \ldots, x_n) \geq u(y_1, \ldots, y_n).$$

[3] We refer to the standard continuity definition. Given two topological spaces (X, τ_X) and (Y, τ_Y), a function $f : X \to Y$ is said to be **continuous (with respect to τ_X and τ_Y)** if for each open subset V of Y, the set $f^{-1}(V)$ is an open subset of X.

If $u : X \rightarrow \mathbb{R}$ is an additive utility function, then for every nonempty set Y and every function $f : Y \rightarrow X$, we have $(u \circ f) = \sum_{i \leq n}(u_i \circ f_i)$, where $f_i : Y \rightarrow X_i$ defined by $f(y) = i\text{-}th\ coordinate\ of\ f(y)$, is the i-th coordinate function of f. Clearly, $(u \circ f)$ satisfies an additive-like property. Thus, abusing notation but in order to be formally consistent, **we introduce the following extension of the notion of additivity** for preference relations defined on a generic nonempty set.

Definition 2.1 *Let $\succsim$ be a preference relation on X. Given a nonempty set Y and a function $f : Y \rightarrow X$, a preference relation can be defined on Y as follows:*

$$\forall y_1, y_2 \in Y,\ y_1 \succsim_f y_2 \overset{def}{\Longleftrightarrow} f(y_1) \succsim f(y_2).$$

*The preference relation $\succsim_f$ will be called the f-***relation induced by** $\succsim$.

Definition 2.2 *Let $\succsim$ be a preference relation on X. Let Y be a nonempty set and fix $f : Y \rightarrow X$. The f-relation $\succsim_f$ will be called* **additive** *if the inducing relation $\succsim$ is additive on X.*

Further topological concepts and standard results will be recall as needed. For a deeper understanding of them, the reader may refer to [15], [19] and [24].

3. Main Assumptions

The following assumptions will hold through the paper.

Assumption 1. For every $i \leq n$, let $\succsim_i$ be a preference relation on X_i and u_i be a bounded (above and below) utility function representing $\succsim_i$.

Henceforth, let $u : X \rightarrow \mathbb{R}$ be defined by:

$$\forall x = (x_1, \ldots, x_n) \in X,\ u(x) = \sum_{i \leq n} u_i(x_i).$$

Since each u_i is an increasing real function, the sum function u is increasing and it induces a preference relation $\succsim_u$ on X, defined as follows:

$$\forall x, y \in X,\ x \succsim_u y \overset{def}{\Longleftrightarrow} u(x) \geq u(y).$$

The preference relation $\succsim_u$ is clearly additive on X.

Assumption 2. Endow X with the preference relation $\succsim_u$.

Henceforth, let $\varphi : \mathcal{G} \rightarrow X$ be defined by $\varphi(G) = (x_1^G, x_2^G, \ldots, x_n^G)$, for every $G \in \mathcal{G}$.

Note that X may contain tuples of characteristics that do not necessarily describe any existing good. Therefore, φ is injective, but not necessarily bijective. Without loss of generality, we will work under the assumption that $X = \varphi(\mathcal{G})$, that is:

Assumption 3. φ is bijective.

Clearly, by Assumption 3, every G in $\mathcal{G}$ corresponds to exactly one n-tuple of X.

Moreover, by means of the map φ, the relation $\succsim_u$ induces the preference relation $\succsim_\varphi$ on $\mathcal{G}$ (see Definition 2.1) which is additive (by Definition 2.2).

Assumption 4. Endow $\mathcal{G}$ with the additive φ-relation $\succsim_\varphi$.

We also assume the decision maker to be endowed with a subjective probability (density) function over each characteristic factor X_i. Abusing notation, each X_i can be considered a random variable.

Assumption 5. For every $i \leq n$, $\mu_i : X_i \rightarrow [0,1]$ is a non-atomic probability density function if X_i is absolutely continuous, and a non-degenerate probability function if X_i is discrete.[4]

Clearly, we do not consider atomic probability density functions or degenerate probability functions, since they do not necessarily induce risk on the choices made by the decision maker.

The functions $\mu_1, \ldots, \mu_n$ must be interpreted as the subjective "beliefs" of the decision maker. For $i \leq n$, $\mu_i(Y_i)$ is the subjective probability that a randomly observed good from $\mathcal{G}$ displays an element $x_i \in Y_i \subseteq X_i$ as its i-th characteristic.[5]

Finally, following the standard economic theory of choice under uncertainty (see [18]), we assume that every decision maker assigns to each unknown i-th

[4]For these and other concepts commonly used in statistical decision theory, see [7].

[5]Note that the functions $\mu_1, \ldots, \mu_n$ can be assumed either independent or correlated, without this fact affecting our results.

characteristic $x_i \in X_i$ the i-th certainty equivalent value induced by her subjective probability (density) function μ_i.

Let $i \leq n$. The **certainty equivalent of** μ_i **and** u_i, denoted by c_i, is a characteristic in X_i that the decision maker is indifferent to accept in place of the expected one to be obtained through (μ_i, u_i).

In other words, for every $i \leq n$, c_i is an element of X_i whose utility $u_i(c_i)$ equals the expected value of u_i. Hence, $c_i \in u_i^{-1}\left(\int_{X_i} u_i(x_i)\mu_i(x_i)dx_i\right)$, if X_i is absolutely continuous, and $c_i \in u_i^{-1}\left(\sum_{x_i \in X_i} u_i(x_i)\mu_i(x_i)\right)$, if X_i is discrete.

The existence of the i-th certainty equivalent characteristic defined by the decision maker in X_i is trivially equivalent to $u_i^{-1}\left(\int_{X_i} u_i(x_i)\mu_i(x_i)dx_i\right)$, or $u_i^{-1}\left(\sum_{x_i \in X_i} u_i(x_i)\mu_i(x_i)\right)$, being a nonempty set. It is not difficult to provide examples of pairs (μ_i, u_i) on the set X_i such that c_i does not exist. In these cases, the decision maker can fix an element of X_i whose utility provides the subjectively closest approximation to the expected value (that is, $\int_{X_i} u_i(x_i)\mu_i(x_i)dx_i$, or $\sum_{x_i \in X_i} u_i(x_i)\mu_i(x_i)$); see [9] and [10]. Clearly, any approximation process generates a bias on the choice of the decision maker. However, as it will become evident below, our results remain unaffected by this fact. Hence, without loss of generality, we will work under the following assumption.

Assumption 6. For every $i \leq n$, c_i exists.

The use of certainty equivalent values implies that if the known characteristic delivers a higher (lower) utility than the corresponding subjective certainty equivalent value, the decision maker prefers the good defined by the former (latter) one.

4. Info-multifunctions and Induced Preferences

Consider a **multifunction** $T : \mathcal{G} \rightrightarrows \{1, 2, \ldots, n\}$, that is, a map that associates to each good G a (possibly empty) finite set of indices. Denote by $Dom(T)$ the domain of T, that is, the set $\{G \in \mathcal{G} : T(G) \neq \emptyset\}$. We interpret each image $T(G)$ as the set of indices corresponding to the known characteristics of the good G. Following this interpretation, a multifunction T becomes a mechanism describing which information and from which good is made available to the decision maker by the information sender.

In a manipulation oriented setting, it is also natural to assume that the information sender would disclose information so as to direct the choice of the decision maker towards a predetermined subset of goods in $Dom(T)$. In order to do so, the information sender cannot allow for the whole information or no information at all to be displayed. Thus, he cannot use as a mechanism the *global info-multifunction* T^{all} defined by:

$$T^{all}(G) = \{1,2,\ldots,n\}, \text{ whenever } G \in \mathcal{G},$$

or the *empty valued info-multifunction* $T_\emptyset$ defined by:

$$T_\emptyset(G) = \emptyset, \text{ whenever } G \in \mathcal{G}.$$

Clearly, $Dom(T^{all}) = \mathcal{G}$. However, requiring $Dom(T) = \mathcal{G}$ for a multifunction $T : \mathcal{G} \rightrightarrows \{1,2,\ldots,n\}$ does not necessarily imply that $T = T^{all}$. Examples of multifunctions $T \neq T^{all}$ such that $Dom(T) = \mathcal{G}$ can be easily given: consider, for instance, $T : \mathcal{G} \rightrightarrows \{1,2,\ldots,n\}$ defined by $T(G) = \{1\}$, whenever $G \in \mathcal{G}$. This idea yields the following definition.

Definition 4.1 *A multifunction* $T : \mathcal{G} \rightrightarrows \{1,2,\ldots,n\}$ *is an* **information multifunction**, *or* **info-multifunction**, *if it is different from both* T^{all} *and* $T_\emptyset$, *and* $Dom(T) \neq \mathcal{G}$. *We will denote the set of all info-multifunctions by* $\mathcal{M}(\mathcal{G}, n)$.

Note that it makes intuitive sense to assume that the goods belonging to $Dom(T)$ provide the decision maker with a higher utility than those in the complementary set. That is,

$$\forall G \in Dom(T), \quad \sum_{i \in T(G)} u_i(x_i^G) > \sum_{i \in T(G)} u_i(c_i).$$

Indeed, the main results presented through the chapter would not be modified if this condition is imposed on the analysis. In the current more general setting, however, we are also allowing the information sender to disclose information on a given set of goods so as to deliver a lower utility to the decision maker than the corresponding sum of certainty equivalent values, i.e. $\sum_{i \in T(G)} u_i(x_i^G) < \sum_{i \in T(G)} u_i(c_i)$. Clearly, these goods would not be considered by the decision maker, but belong to the domain of the info-multifunction.

Furthermore, according to our interpretation, any set of information is induced by a multifunction in $\mathcal{M}(\mathcal{G},n)$, and vice versa. Assigning a multifunction, however, is more general than assigning an information set, since the multifunction does not specify the value of each of the known characteristics. The value of each of the known characteristics remains specified by means of the info-map determined by the given info-multifunction, and defined as follows.

Definition 4.2 *Let $T \in \mathcal{M}(\mathcal{G},n)$. For every $i \leq n$, let $\psi_i^T : \mathcal{G} \to X_i$ be defined by*

$$\psi_i^T(G) = \begin{cases} x_i^G & \text{if } i \in T(G), \\ c_i & \text{otherwise,} \end{cases}$$

with c_i being the i-th certainty equivalent of μ_i and u_i. The product function $\prod_{i \leq n} \psi_i^T : \mathcal{G} \to X$ defined by

$$\left(\prod_{i \leq n} \psi_i^T\right)(G) = (\psi_1^T(G), \ldots, \psi_n^T(G)),$$

where $G \in \mathcal{G}$, and denoted by ψ^T, is the **info-map determined by** *T.*

Given $T \in \mathcal{M}(\mathcal{G},n)$, the info-map ψ^T allows to describe each good as an n-tuple where all unknown characteristics are substituted by their corresponding certainty equivalent values. See Figure 1. Clearly, the info-map ψ^T is not necessarily bijective.[6]

The incomplete information sets encoded in T forces the decision maker to change her original preference relation, $\succsim_\varphi$, and base her choice on the ψ^T-relation induced by $\succsim_u$. See Definition 2.2.

More precisely, for every $G, H \in \mathcal{G}$,

$$G \succsim_{\psi^T} H \overset{def}{\Longleftrightarrow} \psi^T(G) \succsim_u \psi^T(H).$$

$$\Longleftrightarrow u(\psi^T(G)) \geq u(\psi^T(H)) \Longleftrightarrow \sum_{i \leq n} u_i(\psi_i^T(G)) \geq \sum_{i \leq n} u_i(\psi_i^T(H)).$$

The following result is now immediate.

[6]If we were to consider them, the info-map $\psi^{T_{all}}$ would be equal to the identification map φ, while the info-map $\psi^{T_\emptyset}$ would be the constant function defined by $\psi^{T_\emptyset}(G) = (c_1, \ldots, c_n)$, whenever $G \in \mathcal{G}$.

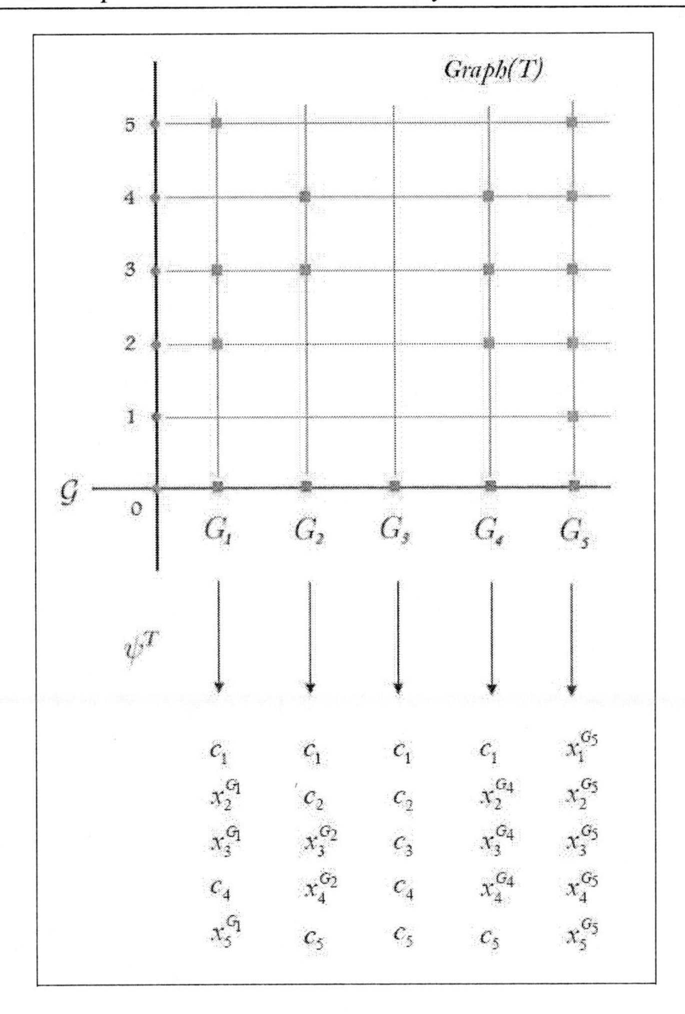

Figure 1. The upper part represents the graph of T whose domain is $Dom(T) = G_1, G_2, G_3, G_4, G_5$. The goods are supposed to be described by a total of $n = 5$ characteristics. The lower part illustrates the info-map determined by T.

Proposition 4.3 *For every $T \in \mathcal{M}(\mathcal{G}, n)$, the preference relation $\succsim_{\psi^T}$ is additive on $\mathcal{G}$ and represented by $u \circ \psi^T$.*

Note that $\succsim_{\psi^T}$ is in general different from $\succsim_{\varphi}$. Therefore, different preference relations can be induced depending on the information set presented to the decision maker. This implies that knowing the original preference relation of a decision maker, $\succsim_{\varphi}$, allows for displaying information sets in such a way so as to manipulate her final choice.

More precisely, an information sender who knows u_i, for $i \leq n$ (equivalently, $\succsim_i$ for $i \leq n$), as well as μ_i, for $i \leq n$, can manipulate the choice made by the decision maker. As already mentioned (see Section 1), this assumption is in line with the common knowledge of utilities and beliefs used by the cited literature to define full revelation mechanisms.

5. Continuously Representable Induced Preferences

We shall now investigate sufficient conditions for a generic info-multifunction to induce continuously representable preference relations.

In order to do so, we need to expand our initial set of assumptions by introducing a topology on each of the sets so far involved in our model. We endow the reals $\mathbb{R}$ with the standard Euclidean topology. Besides, we add the following assumptions to those introduced in Section 3.

Assumption 7. For every $i \leq n$, let τ_i and $\succsim_i$ be a connected[7] topology and a preference relation on X_i such that there exists a continuous utility function $u_i : (X_i, \tau_i) \to \mathbb{R}$ representing $\succsim_i$.

Assumption 7 is clearly stronger than Assumption 1, since the latter is implied by the former.

Assumption 8. Let X be endowed with the product topology τ_p and assume the utility function $u = u_1 + u_2 + \cdots + u_n : (X, \tau_p) \to \mathbb{R}$ to be continuous.

Assumption 9. Let $\mathcal{G}$ be endowed with the weak topology τ_{φ} induced by the function $\varphi : \mathcal{G} \to (X, \tau_p)$.

Recall that, given a nonempty set Y, a topological space (Z, τ_Z) and a function $f : Y \to Z$, the **weak topology on Y determined by a function f** is the topology having as a subbase all sets of the form $f^{-1}(V)$, where V is open in Z.

[7] A topological space is *connected* if there are no two disjoint nonempty open subsets (or equivalently, two nonempty disjoint closed subsets) whose union equals the space; and *disconnected* if it is not connected.

It is easy to check that the weak topology on a set Y induced by a function f is the weakest topology with respect to which the function f is continuous.[8]

Since the product of connected spaces is connected, and the closed subsets of $\mathcal{G}$ are preimages via φ of closed subsets of X, the new assumptions imply the connectedness of the space $(\mathcal{G}, \tau_\varphi)$.

For a discussion concerning the conditions sufficient for Assumption 7 and Assumption 8 to be verified we refer the reader to [5], [6], [8].

The new set of assumptions leads to the following results.

Proposition 5.1 *For every $i \leq n$, the function $\varphi_i : \mathcal{G} \rightarrow X_i$ defined by $\varphi_i(G) = x_i^G$ is continuous.*

Proof: Note that $\varphi = \prod_{i \leq n} \varphi_i$. Since, φ is continuous (Assumption 9), each φ_i is continuous.[9] ∎

Proposition 5.2 *The preference relation $\succsim_\varphi$ is additive and continuously representable on $\mathcal{G}$.*

Proof: By Assumptions 8 and 9, $u \circ \varphi$ is a continuous function from $\mathcal{G}$, endowed with τ_φ, to the reals (endowed with the standard topology). Clearly, $u \circ \varphi$ represents $\succsim_\varphi$ and its additivity follows from Assumption 4. ∎

Let $T \in \mathcal{M}(\mathcal{G}, n)$. If $Dom(T)$ is discrete and finite, it is straightforward to show that the restriction of ψ^T to $Dom(T)$, denoted by $\psi^T \upharpoonright_{Dom(T)}$, is continuous. Hence, by Proposition 4.3, $u \circ \left(\psi^T \upharpoonright_{Dom(T)}\right)$ is an additive continuous utility function representing $\succsim_{\psi^T}$ over $Dom(T)$.

We shall show that the restriction $\psi^T \upharpoonright_{Dom(T)}$ can be proved to be continuous even if we relax the finiteness assumption on the set $Dom(T)$.

Given a $T \in \mathcal{M}(\mathcal{G}, n)$, we can define an equivalence relation on the corresponding domain $Dom(T)$ as follows.

---

[8]More in general, a topology can be determined by a family of functions (see Section 1.2 in [2]). Let $\{Z_\alpha : \alpha \in \Lambda\}$ be a family of topological spaces and let Y be a nonempty set. Suppose that $\mathcal{F} = \{f_\alpha : \alpha \in \Lambda\}$ is a family of functions, where for every α, $f_\alpha : Y \rightarrow Z_\alpha$. The **weak topology on X determined by $\mathcal{F}$** is the topology having as a subbase all sets of the form $f_\alpha^{-1}(V_\alpha)$, where V_α is open in Z_α.

[9]Let $\prod_{\alpha \in J} Z_\alpha$ be product of topological spaces endowed with the product topology and Y be a topological space. Let $f : Y \rightarrow \prod_{\alpha \in J} Z_\alpha$ be given by the equation $f(y) = (f_\alpha(y))_{\alpha \in J}$, where $f_\alpha : Y \rightarrow Z_\alpha$ for each α. Then, the function f is continuous if and only if each function f_α is continuous. See Theorem 19.6 in [24].

Definition 5.3 *Let $T \in \mathcal{M}(\mathcal{G}, n)$ and fix $G, H \in Dom(T)$. We will say that G is T-equivalent to H, and write $G \lhd_T \rhd H$, if $T(G) = T(H)$.*

The fact that $\lhd_T \rhd$ defines an equivalence relation on $Dom(T)$ is trivial. The equivalence class determined by $G \in Dom(T)$ is the set $[G]_{\lhd_T \rhd} = \{H \in Dom(T) : T(H) = T(G)\}$.

In order to simplify notations, we will let $\Delta(T)$ stand for the quotient set $Dom(T)/ \lhd_T \rhd$ and $\mathcal{D}$ denote the generic equivalence class. Finally, we will use $T(\mathcal{D})$ to denote each of the sets $T(G)$, where G varies in $\mathcal{D}$.[10] Thus, $T(\mathcal{D})$ will denote the set of the indexes corresponding to all the known characteristics of each of the goods in $\mathcal{D}$.

Being a quotient set, the family $\Delta(T)$ is a partition of $Dom(T)$. Moreover, since for every $G \in Dom(T)$, $T(G) \subseteq \{1, 2, \ldots, n\}$, we have that for every $\mathcal{D} \in \Delta(T)$, $T(\mathcal{D}) \subseteq \{1, 2, \ldots, n\}$. Consequently, $\Delta(T)$ has cardinality at most 2^n.

Thus, to each info-multifunction T remains associated a unique finite partition of $Dom(T)$.

Lemma 5.4 *Let $T \in \mathcal{M}(\mathcal{G}, n)$. Then, for every $\mathcal{D} \in \Delta(T)$, $\psi^T \restriction_{\mathcal{D}}$ is continuous.*

Proof: Fix $\mathcal{D} \in \Delta(T)$ and note that $\psi^T \restriction_{\mathcal{D}} = \prod_{i \leq n} \psi_i^T \restriction_{\mathcal{D}}$.

If $i \in T(\mathcal{D})$, then $\psi_i^T \restriction_{\mathcal{D}} = \varphi_i \restriction_{\mathcal{D}}$. If $i \notin T(\mathcal{D})$, then $\psi_i^T \restriction_{\mathcal{D}}$ is the constant function $H \in \mathcal{D} \to c_i \in X_i$. See Figure 2.

Consequently, $\psi^T \restriction_{\mathcal{D}}$ is the product of continuous functions; hence, it is continuous.[11] ∎

Lemma 5.5 *Let $T \in \mathcal{M}(\mathcal{G}, n)$. If, for every $\mathcal{D} \in \Delta(T)$:*

(a) $\mathcal{D}$ is a closed subset of $\mathcal{G}$;

(b) $\psi^T \restriction_{\mathcal{D}}$ is continuous;

then, the map $\psi^T \restriction_{Dom(T)}$ is continuous.

Proof: Apply the so-called pasting lemma (Theorem 18.3 and Exercise 18.9 in [24]). ∎

[10] In fact, all sets $T(G)$ are equal when G varies in $\mathcal{D}$.

[11] By Theorem 19.6 in [24]. See also footnote 9.

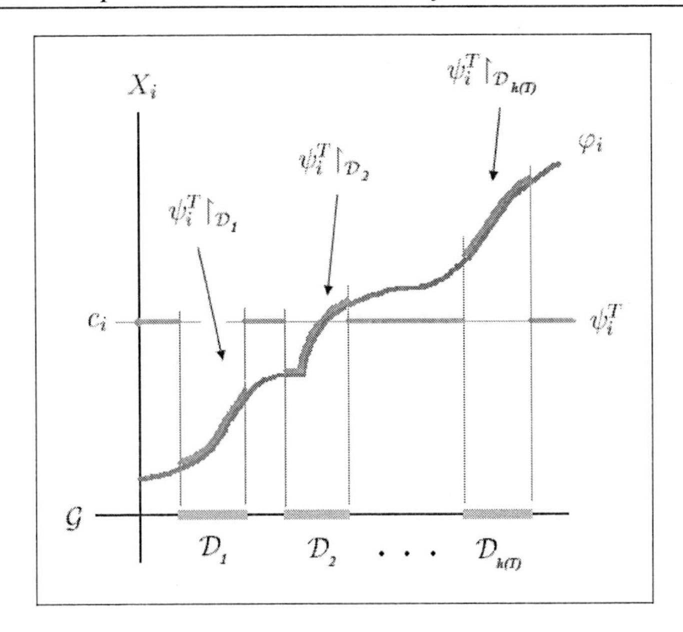

Figure 2. A graphical representation of the proof of Lemma 5.4 with respect to the single i-th characteristic factor space Xi in the particular case when $n = 5$ and the equivalence classes are connected subsets. Note that h(T) denotes the cardinality of $\delta(T)$.

Note that condition (a) of Lemma 5.5. can hold true for an info-multifunction T thanks to the requirement $Dom(T) \neq \mathcal{G}$. Allowing for $Dom(T) = \mathcal{G}$ would imply that $\mathcal{G}$ is disconnected, contradicting Assumptions 7-9.

Theorem 5.6 *Let $T \in \mathcal{M}(\mathcal{G}, n)$. If $\Delta(T)$ consists of closed subsets of $\mathcal{G}$, then the preference relation $\succsim_{\psi^T}$ is additive and continuously represented on $Dom(T)$ by $u \circ \psi^T \lceil_{Dom(T)}$.*

Proof: By Proposition 4.3, the preference relation $\succsim_{\psi^T}$ is additive on $\mathcal{G}$ and represented by $u \circ \psi^T$. At the same time, by Lemmas 5.4 and 5.5, $\psi^T \lceil_{Dom(T)}$ is continuous. Hence, $u \circ \psi^T \lceil_{Dom(T)}$ is continuous. ∎

Henceforth, we will denote by $\mathcal{M}^*(\mathcal{G}, n)$ the set of all info-multifunctions T whose associated partition $\Delta(T)$ consists of closed subsets of $\mathcal{G}$.

6. Defining Homotopic Preferences

For $T \in \mathcal{M}(\mathcal{G}, n)$, the graph of T is the set $Graph(T) = \{(G,i) : G \in Dom(T) \wedge i \in T(G)\}$. We will consider the following partial order on the set of all info-multifunctions on $\mathcal{G}$.[12]

Definition 6.1 *Let $T_1, T_2 \in \mathcal{M}(\mathcal{G}, n)$. We say that T_1 is a* **shrinking** *of T_2, or that T_2 is a* **dilatation** *of T_1, if $Graph(T_1) \subseteq Graph(T_2)$ (equivalently, $\forall G \in \mathcal{G}$, $T_1(G) \subseteq T_2(G)$).*

The shrinking and dilatation concepts defined above can be respectively associated to the loss and acquisition of information processes. While the latter concept is reasonable, as decision makers tend to receive new information through time, the former one may seem counterintuitive. However, it can be theoretically justified using the concept of limited memory. This notion is widely used in economic theory to model environments where decision makers display a limited capacity to assimilate and remember information, see for example [12].

What we show with the use of basic homotopic results is that continuously representable preference relations determined by dilatations or shrinkings of a given info-multifunction can be continuously transformed one into another. This property implies that *an informed sender is able to manipulate the preference relation and, consequently, the choice of a decision maker through time and in a smooth way.*

Let T be given at time $t = 0$ and suppose that after a (continuous) time interval, the sender releases further information encoded in a dilation of T denoted by $\mathbb{D}$. If both $u \circ \psi^T$ and $u \circ \psi^{\mathbb{D}}$ are continuous on a certain subset $\mathcal{H}$ of $\mathcal{G}$, then the map $F^d : \mathcal{H} \times [0,1] \to \mathbb{R}$, defined by $F^d(G,t) = (1-t) \cdot u(\psi^T(G)) + t \cdot u(\psi^{\mathbb{D}}(G))$, is a homotopy continuously transforming $u \circ \psi^T$ in $u \circ \psi^{\mathbb{D}}$.[13]

Similarly, if the sender decides to reduce the initial information encoded in T using a shrinking $\mathbb{S}$ such that $u \circ \psi^T$ and $u \circ \psi^{\mathbb{S}}$ are both continuous on a given subset $\mathcal{H}$ of $\mathcal{G}$, then there exists an homotopy $F^s : \mathcal{H} \times [0,1] \to \mathbb{R}$ continuously transforming $u \circ \psi^T$ in $u \circ \psi^{\mathbb{S}}$.

---

[12]Definition 6.1 is an interpretation in information terms of the standard definition of submultifunction/supermultifunction. See for example Section 6.1 in [2].

[13]The homotopy F^d is known as the **straight line homotopy** and can be defined between any two continuous functions taking values in a convex subsets of $\mathbb{R}^n$. See also Section 51 in [24].

Consequently, Theorem 5.6, stating the conditions sufficient for $\psi^T \lceil_{Dom(T)}$ to be continuous, becomes the key result of the paper. Indeed, Theorem 5.6 can be used to show that a necessary condition for the homotopy F^d (resp. F^s) to exist is that $Dom(\mathbb{D}) = Dom(T)$ (resp. $Dom(\mathbb{S}) = Dom(T)$).

In other words, in order to be able to define a homotopy between the initially induced preference relation $\succsim_{\psi^T}$ and the new one $\succsim_{\psi^{\mathbb{D}}}$ (resp. $\succsim_{\psi^{\mathbb{S}}}$) resulting from dilatating (resp. shrinking) the information transmitted, both these preference relations must be defined and continuously representable on the same subset of $\mathcal{G}$, namely $Dom(T)$.

These remarks lead to the introduction of the following notion of *compatibility*.

Definition 6.2 *Let $T \in \mathcal{M}(\mathcal{G}, n)$. A dilatation $\mathbb{D}$ (resp. shrinking $\mathbb{S}$) of T is called* **compatible** *if $Dom(\mathbb{D}) = Dom(T)$ (resp. $Dom(\mathbb{S}) = Dom(T)$).*

For every $T \in \mathcal{M}(\mathcal{G}, n)$, let $\mathcal{CD}(T)$ denote the set of all compatible dilatations of T, and $\mathcal{CS}(T)$ denote the set of all compatible shrinkings of T.

Note that examples of info multifunctions that do not admit any compatible dilatation or shrinking can be easily constructed.

Consider, for instance, the info-multifunction $T \in \mathcal{M}(\mathcal{G}, n)$ defined as follows:

$$T(\widetilde{G}) = \{1, \ldots, n\} \text{ and } T(G) = \emptyset, \ \forall G \neq \widetilde{G},$$

where $\widetilde{G}$ is a fixed good in $\mathcal{G}$. It is easy to check that $Dom(T) = \{\widetilde{G}\}$, while $\mathcal{CD}(T) = \emptyset$. See Figure 3(a).

An example of info-multifunction T such that $\mathcal{CS}(T) = \emptyset$ is the following:

$$T(\widetilde{G}) = \{1\} \text{ and } T(G) = \emptyset, \ \forall G \neq \widetilde{G}$$

where $\widetilde{G}$ is again a fixed good in $\mathcal{G}$. See Figure 3(b).

To simplify the presentation of the results, we extend the homotopic relationship between the utility functions representing the preference relations induced by T and by $\mathbb{D} \in \mathcal{CD}(T)$ (resp. $\mathbb{S} \in \mathcal{CS}(T)$) to the preference relations themselves.

Definition 6.3 *Let $T \in \mathcal{M}(\mathcal{G}, n)$. For every $\mathbb{D} \in \mathcal{CD}(T)$, the preference relations $\succsim_{\psi^T}$ and $\succsim_{\psi^{\mathbb{D}}}$ will be called* **homotopic** *if the utility functions $u \circ \psi^T$*

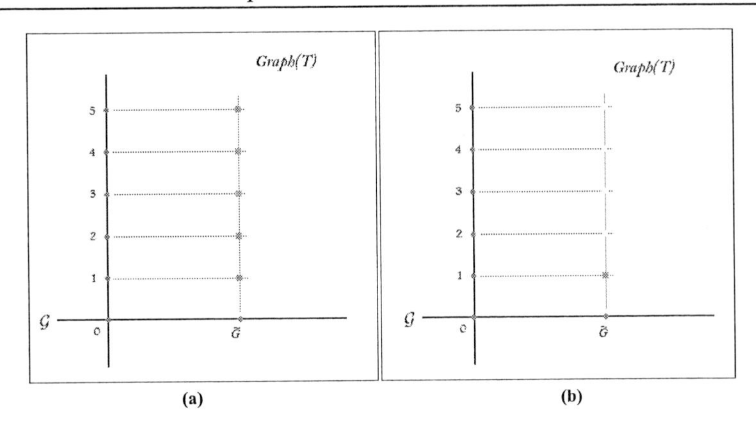

Figure 3.

and $u \circ \psi^{\mathbb{D}}$ *are continuous and homotopic on* $Dom(T)$. *Similarly, for every* $\mathbb{S} \in \mathcal{CS}(T)$, *the preference relations* $\succsim_{\psi^T}$ *and* $\succsim_{\psi^{\mathbb{S}}}$ *will be called* **homotopic** *if the utility functions* $u \circ \psi^T$ *and* $u \circ \psi^{\mathbb{S}}$ *are continuous and homotopic on* $Dom(T)$.

Theorem 6.4 *Let* $T \in \mathcal{M}(\mathcal{G}, n)$.

(1) *The following are equivalent:*

(1.a) $\mathcal{CD}(T) \neq \emptyset$;

(1.b) $\exists G \in Dom(T)$ *such that* $|T(G)| < n$;

(1.c) $\exists \widetilde{\mathcal{D}} \in \Delta(T)$ *such that* $|T(\widetilde{\mathcal{D}})| < n$ *(that is,* $\exists j \leq n$ *such that the j-th characteristic is still unknown for all goods belonging to* $\widetilde{\mathcal{D}}$*).*

(2) *The following are equivalent:*

(2.a) $\mathcal{CS}(T) \neq \emptyset$;

(2.b) $\exists G \in Dom(T)$ *such that* $|T(G)| \geq 2$;

(2.c) $\exists \widetilde{\mathcal{D}} \in \Delta(T)$ *such that* $|T(\widetilde{\mathcal{D}})| \geq 2$ *(that is,* $\exists j \leq n$ *such that the j-th characteristic is already known for all goods belonging to* $\widetilde{\mathcal{D}}$*, but it is not the only one).*

Theorem 6.5 *Let* $T \in \mathcal{M}^{\star}(\mathcal{G}, n)$.

(1) *If* $\mathbb{D} \in \mathcal{CD}(T) \cap \mathcal{M}^\star(\mathcal{G}, n)$, *then* $\succsim_{\psi^{\mathbb{D}}}$ *is homotopic to* $\succsim_{\psi^T}$ *on* $Dom(T)$.

(2) *If* $\mathbb{S} \in \mathcal{CS}(T) \cap \mathcal{M}^\star(\mathcal{G}, n)$, *then* $\succsim_{\psi^{\mathbb{S}}}$ *is homotopic to* $\succsim_{\psi^T}$ *on* $Dom(T)$.

Proof. Part (1). By Theorem 5.6, $u \circ \psi^T \restriction_{Dom(T)}$ and $u \circ \psi^{\mathbb{D}} \restriction_{Dom(\mathbb{D})}$ are both continuous. Since $Dom(T) = Dom(\mathbb{D})$, they are homotopic by means of the straight line homotopy $F : Dom(T) \times [0,1] \to \mathbb{R}$ defined by $F(G,t) = (1-t) \cdot u(\psi^T(G)) + t \cdot u(\psi^{\mathbb{D}}(G))$.

Part (2). Similar to the proof of part (1). ■

7. Uniformity and Homotopic Preferences

Let $T \in \mathcal{M}(\mathcal{G}, n)$ be such that $\mathcal{CD}(T) \neq \emptyset$. For every $\mathcal{D} \in \Delta(T)$, let $J_{\mathcal{D}} = \{1, \ldots, n\} \backslash T(\mathcal{D})$.

$J_{\mathcal{D}}$ is the set of the indexes corresponding to the characteristics which are still unknown for all the goods in $\mathcal{D}$. Clearly, $J_{\mathcal{D}} = \emptyset$ if $T(\mathcal{D}) = \{1, \ldots, n\}$.

Now, let $\Gamma(T) = \{\mathcal{D} \in \Delta(T) : J_{\mathcal{D}} \neq \emptyset\}$. Note that $\Gamma(T) \neq \emptyset$ since condition $(1.c)$ of Theorem 6.4 holds.

For every family $\{I_{\mathcal{D}} : \mathcal{D} \in \Theta\}$, where $\Theta \subseteq \Gamma(T)$ and, for every $\mathcal{D} \in \Theta$, $I_{\mathcal{D}} \subseteq J_{\mathcal{D}}$, consider the info-multifunction $\mathbb{D}^{\{I_{\mathcal{D}} : \mathcal{D} \in \Theta\}}$ defined by:

$$\mathbb{D}^{\{I_{\mathcal{D}} : \mathcal{D} \in \Theta\}}(G) = \begin{cases} T(G) \cup I_{\mathcal{D}} & \text{if } G \in \mathcal{D}, \\ T(G) & \text{if } G \notin \bigcup_{\mathcal{D} \in \Theta} \mathcal{D}. \end{cases}$$

It is easy to check that $\mathbb{D}^{\{I_{\mathcal{D}} : \mathcal{D} \in \Theta\}}$ is a compatible dilatation of T. Given its uniform character on each one of the classes of equivalence associated to T, we can classify such a dilatation as "uniform".

Definition 7.1 *Let* $T \in \mathcal{M}(\mathcal{G}, n)$ *be such that* $\mathcal{CD} \neq \emptyset$. *A compatible dilatation of* T *will be called the* **uniform** *if it is of the form* $\mathbb{D}^{\{I_{\mathcal{D}} : \mathcal{D} \in \Theta\}}$.

It is also clear that uniform dilatations of T can be defined as soon as the compatibility condition holds. More precisely, we can state the following.

Proposition 7.2 *Let* $T \in \mathcal{M}(\mathcal{G}, n)$. *If* $\mathcal{CD}(T) \neq \emptyset$, *then* T *admits at least one uniform dilatation.*

Example 7.3 *Suppose that each good in $\mathcal{G}$ is described by $n = 5$ characteristics and fix three disjoint subsets of $\mathcal{G}$, $\mathcal{D}_1$, $\mathcal{D}_2$ and $\mathcal{D}_3$.*

Let $T \in \mathcal{M}(\mathcal{G}, 5)$ be defined as follows:

$$T(G) = \begin{cases} \{4\} & \text{if } G \in \mathcal{D}, \\ \{3,4\} & \text{if } G \in \mathcal{D}, \\ \{1,2,5\} & \text{if } G \in \mathcal{D}, \\ \emptyset & \text{if } G \notin \mathcal{D}_1 \cup \mathcal{D}_2 \cup \mathcal{D}_3. \end{cases}$$

Clearly, $Dom(T) = \mathcal{D}_1 \cup \mathcal{D}_2 \cup \mathcal{D}_3$ and $\Delta(T) = \{\mathcal{D}_1, \mathcal{D}_2, \mathcal{D}_3\}$. See Figure 4(a). We can compatibly dilatate the information encoded in T by expanding the information known for the goods in $\mathcal{D}_1$, $\mathcal{D}_2$ and $\mathcal{D}_3$ as follows. Add the information on the 2-nd, 3-rd and 5-th characteristics for all the goods in $\mathcal{D}_1$, on the 5-th characteristic for all the goods in $\mathcal{D}_2$, and on the 3-rd and 4-th characteristics for all the goods in $\mathcal{D}_3$. This leads to the dilatation $\mathbb{D}^{\{I_{\mathcal{D}} : \mathcal{D} \in \Theta\}}$ determined by the family $\{I_{\mathcal{D}_1}, I_{\mathcal{D}_2}, I_{\mathcal{D}_3}\}$, where $I_{\mathcal{D}_1} = \{2,3,5\}$, $I_{\mathcal{D}_2} = \{5\}$ and $I_{\mathcal{D}_3} = \{3,4\}$. See Figure 4(b).

Following a reasoning dual to the one developed above, we can state the existence of uniform shrinkings for info-multifunctions that admit at least one compatible shrinking.

Let $T \in \mathcal{M}(\mathcal{G}, n)$ be such that $\mathcal{CS} \neq \emptyset$. Let $\Gamma(T) = \{\mathcal{D} \in \Delta(T) : |T(\mathcal{D})| \geq 2\}$. Note that $\Gamma(T) \neq \emptyset$ since condition (2.c) of Theorem 6.4 holds.

For every family $\{I_{\mathcal{D}} : \mathcal{D} \in \Theta\}$, where $\Theta \subseteq \Gamma(T)$ and, for every $\mathcal{D} \in \Theta$, $I_{\mathcal{D}}$ is a proper subset of $T(\mathcal{D})$, consider the info-multifunction $\mathbb{S}^{\{I_{\mathcal{D}} : \mathcal{D} \in \Theta\}}$ defined by:

$$\mathbb{S}^{\{I_{\mathcal{D}} : \mathcal{D} \in \Theta\}}(G) = \begin{cases} T(G) \setminus I_{\mathcal{D}} & \text{if } G \in \mathcal{D}, \\ T(G) & \text{if } G \notin \bigcup_{\mathcal{D} \in \Theta} \mathcal{D}. \end{cases}$$

By construction, $\mathbb{S}^{\{I_{\mathcal{D}} : \mathcal{D} \in \Theta\}}$ is a compatible shrinking of T, and it is uniform.

Definition 7.4 *Let $T \in \mathcal{M}(\mathcal{G}, n)$ be such that $\mathcal{CS} \neq \emptyset$. A compatible shrinking of T will be called the* **uniform** *if it is of the form $\mathbb{S}^{\{I_{\mathcal{D}} : \mathcal{D} \in \Theta\}}$.*

Proposition 7.5 *Let $T \in \mathcal{M}(\mathcal{G}, n)$. If $\mathcal{CS}(T) \neq \emptyset$, then T admits at least one uniform shrinking.*

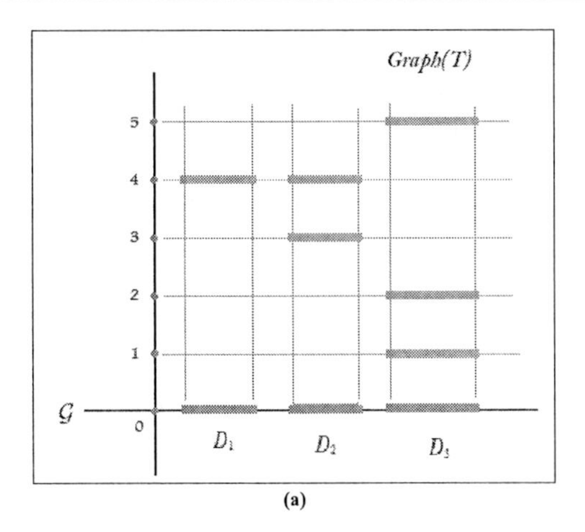

(a)

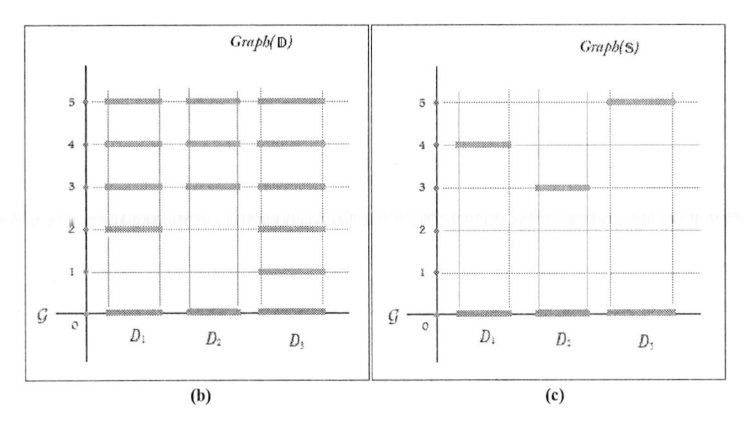

(b) (c)

Figure 4.

Example 7.6 *Suppose that each good in $\mathcal{G}$ is described by $n = 5$ characteristics and fix three disjoint subsets of $\mathcal{G}$, $\mathcal{D}_1$, $\mathcal{D}_2$ and $\mathcal{D}_3$.*

Let $T \in \mathcal{M}(\mathcal{G}, 5)$ be the multifunction of Example 7.3. We can compatibly shrink the information encoded in T by restricting the information known for the goods in $\mathcal{D}_2$ and $\mathcal{D}_3$ in the following way. Cancel the information on the 4-th characteristic for all the goods in $\mathcal{D}_2$, and on the 1-st and 2-nd characteristics for all the goods in $\mathcal{D}_3$. We obtain the shrinking $\mathbb{S}^{\{I_{\mathcal{D}}:\mathcal{D}\in\Theta\}}$ determined by the family $\{I_{\mathcal{D}_2}, I_{\mathcal{D}_3}\}$, where $I_{\mathcal{D}_2} = \{4\}$ and $I_{\mathcal{D}_3} = \{1,2\}$. See Figure 4(c).

Note that no compatible shrinking can be obtained by cancelling information on the goods in $\mathcal{D}_1$.

The following result states that the property of admitting a partition of the domain consisting of closed subsets is preserved by all the uniform dilatations and shrinkings of an info-multifunction.

Proposition 7.7 *Let $T \in \mathcal{M}^\star(\mathcal{G},n)$ be such that $\mathcal{CD}(T) \neq \emptyset$. Then, all uniform dilatations and uniform shrinkings of T belong to $\mathcal{M}^\star(\mathcal{G},n)$.*

Proof. By Proposition 7.2, T admits at least one uniform dilatation. Let $\mathbb{D}$ be the uniform dilatation of T determined by the family $\{I_{\mathcal{D}} : \mathcal{D} \in \Theta\}$. Since a uniform dilatation is compatible (see Definition 7.1), $Dom(\mathbb{D}) = Dom(T)$. At the same time, only one of the following holds:

Case 1: $\forall \mathcal{D} \in \Theta, \forall \mathcal{B} \in \Delta(T) \backslash \Theta, \quad T(\mathcal{D}) \cup I_{\mathcal{D}} \neq T(\mathcal{B})$.

Case 2: $\exists \widetilde{\mathcal{D}} \in \Theta, \exists \widetilde{\mathcal{B}} \in \Delta(T) \backslash \Theta,$ such that $T(\widetilde{\mathcal{D}}) \cup I_{\widetilde{\mathcal{D}}} = T(\widetilde{\mathcal{B}})$.

In the first case, the quotient sets determined by the equivalence relations $\lhd_{\mathbb{D}} \rhd$ and $\lhd_T \rhd$ coincide, that is, $\Delta(\mathbb{D}) = \Delta(T)$.

In the second case, $\Delta(\mathbb{D}) \neq \Delta(T)$. The equivalence classes of $\lhd_{\mathbb{D}} \rhd$ are all unions of the form $\mathcal{D} \cup \bigcup \{\mathcal{B} \in \Delta(T) \backslash \Theta : T(\mathcal{D}) \cup I_{\mathcal{D}} = T(\mathcal{B})\}$.

In both cases, the family $\Delta(\mathbb{D})$ consists of closed subsets of $\mathcal{G}$, since this is the case for $\Delta(T)$ by assumption. Consequently, $\mathbb{D} \in \mathcal{M}^\star(\mathcal{G},n)$.

The proof that a generic shrinking of T still belongs to $\mathcal{M}^\star(\mathcal{G},n)$ is similar. ∎

Combined with Theorem 6.5, Proposition 7.7 allows us to conclude that an information sender can manipulate in a continuous way and through a fixed time interval the preferences of a decision maker via uniform dilatations or shrinkings of an initially assigned info-multifunction.

Theorem 7.8 *Let $T \in \mathcal{M}^\star(\mathcal{G},n)$.*

(a) *For every uniform dilatation $\mathbb{D}$ of T, $\precsim_{\psi^{\mathbb{D}}}$ is homotopic to $\precsim_{\psi^T}$ on $Dom(T)$.*

(b) *For every uniform shrinking $\mathbb{S}$ of T, $\precsim_{\psi^{\mathbb{S}}}$ is homotopic to $\precsim_{\psi^T}$ on $Dom(T)$.*

Proof. By Proposition 7.7 and Theorem 6.5. ∎

8. Conclusion

The present paper has provided an innovative interpretation of the idea of multifunction in economic theoretical terms. We believe that this set-theoretical tool, barely used by economic theorists, may be successfully applied in decision theory. We refer the interested reader to the seminal papers of Michael ([20], [21], [22], [23]) and the subsequent literature on multifunctions and continuous selection problems. A brief literature review on the main use of multifunctions in set-theoretic topology and some extensions within this research field can be found in [11].

At the same time, homotopy theory has been used to illustrate how, despite the assumed verifiability of all the information transmitted by a sender, the preferences of a decision maker can be manipulated through time in a continuous way so as to induce any a priori assigned preference relation. Even though the homotopy techniques employed may not be complex in themselves, it is the design of a theoretical economic decision scenario allowing for their application what deserves particular attention. In this regard, we have aimed at extending the limits of economic theory in dealing with complex theoretical structures not explicitly defined by economists. On the other hand, new applications of homotopy theory in economics, not yet considered by mathematicians, have been introduced.

References

[1] M. Battaglini, Multiple Referrals and Multidimensional Cheap Talk, *Econometrica*, **70** (2002), 1379 - 1401.

[2] G. Beer, *Topologies on Closed and Closed Convex Sets*, Kluwer Academic Publishers, 1993.

[3] A. Chakraborty and R. Harbaugh, Comparative Cheap Talk, *Journal of Economic Theory*, **132** (2007), 70 - 94.

[4] V. Crawford and J. Sobel, Strategic Information Transmission, *Econometrica*, **50** (1982), 1431 - 1451.

[5] G. Debreu, Representation of a preference ordering by a numerical function (1954). In: G. Debreu, *Mathematical economics: Twenty papers of Gerard Debreu*, Cambridge University Press, 1983.

[6] G. Debreu, Topological methods in cardinal utility theory (1960). In: G. Debreu, *Mathematical economics: Twenty papers of Gerard Debreu*, Cambridge University Press, 1983.

[7] M. DeGroot, *Optimal Statistical Decisions*, John Wiley & Sons, New Jersey, 2004.

[8] D. Di Caprio and F. Santos-Arteaga, Continuous Lexicographic Choice Through Incomplete Information, *Working Paper* no. **52**, Free University of Bozen-Bolzano, 2007.

[9] D. Di Caprio and F. Santos-Arteaga, Rationally Induced Choice Errors: Error Multifunctions and Generalized Certainty Equivalents, *Working Paper* no. **53**, Free University of Bozen-Bolzano, 2007.

[10] D. Di Caprio and F. Santos-Arteaga, Error-Induced Certainty Equivalents: A Set-Theoretical Approach to Choice under Risk, *International Journal of Contemporary Mathematical Sciences*, **3** (2008), 1121 - 1131.

[11] D. Di Caprio and S. Watson, Continuous Selections and Purely Topological Convex Structures, *Topology Proceedings*, **29** (2005), 75 - 103.

[12] J. Dow, Search Decisions with Limited Memory, *Review of Economic Studies*, **58** (1991), 1 - 14.

[13] U. Dulleck and R. Kerschbamer, On Doctors, Mechanics, and Computer Specialists: The Economics of Credence Goods, *Journal of Economic Literature*, **XLIV** (2006), 5-42.

[14] B.C. Eaves and K. Schmedders, General Equilibrium Models and Homotopy Methods, *Journal of Economic Dynamics & Control*, **23** (1999), 1249 - 1279.

[15] R. Engelking, *General Topology*, Heldermann Verlag, Berlin, 1989.

[16] W. Huffman, M. Rousu, J. Shogren and A. Tegene, The Effects of Prior Beliefs and Learning on Consumers' Acceptance of Genetically Modified Foods, *Journal of Economic Behavior & Organization*, **63** (2007), 193 - 206.

[17] L. Lauwers, Topological Social Choice, *Mathematical Social Sciences*, **40** (2000), 1-39.

[18] A. Mas-Colell, M.D. Whinston and J.R. Green, *Microeconomic Theory*, Oxford University Press, New York, 1995.

[19] W.S. Massey, *A Basic Course in Algebraic Topology*, Springer-Verlag, New York, 1991.

[20] E. Michael, Continuous selections I, *Ann. Math.*, **63** (1956), 361–382.

[21] E. Michael, Continuous selections II, *Ann. Math.*, **64** (1956), 562–580.

[22] E. Michael, Continuous selections III, *Ann. Math.*, **65** (1957), 375–390.

[23] E. Michael, Convex Structures and Continuous Selections, *Can. J. Math.*, **11** (1959), 556–575.

[24] J.R. Munkres, *Topology*, Prentice Hall, Inc, 2000.

[25] P. Wakker, Continuity of preference relations for separable topologies, *International Economic Review*, **29** (1988), 105 - 110.

[26] P. Wakker, *Additive Representations of Preferences, A New Foundation of Decision Analysis*, Dordrecht, Kluwer Academic Publishers, 1989.

In: Perspectives in Applied Mathematics
Editor: Jordan I. Campbell, pp. 117-133
ISBN: 978-1-61122-796-3

Chapter 7

ANALYSIS OF A HETEROGENEOUS FORAGING SWARM MODEL INSPIRED BY HONEYBEE BEHAVIOR

Jia Song[1,2,a], Shengwei Yu[3,b] and Li Xu[1,c]

[1] College of Electrical Engineering, Zhejiang University, P.R. China
[2] Department of Electronic Engineering, Suzhou Vocational College, P.R. China
[3] School of Electrical and Computer Engineering, Purdue University, U.S.A

Abstract

Inspired by the organized behaviors of honeybee swarms, an individual-based mathematical model is proposed in this paper for the heterogeneous swarm. The heterogeneous swarm is assumed to consist of two different kinds of individuals, namely, the scouts and the normal agents, with respect to their sensing abilities. Besides, a short-distance-bounded-attraction function was proposed to describe the attraction among individuals.

Firstly the heterogeneous swarm model is identified and the swarm cohesion is proved, and the analytical bound on the swarm size is provided. Secondly, the foraging properties of the heterogeneous swarm in multimodal Gaussian environment are studied, and conditions for collective convergence to more favorable regions are provided. Thirdly, simulations were carried out and the

[a] E-mail address: sjia@jssvc.edu.cn
[b] E-mail address: shinewaysw@hotmail.com
[c] E-mail address: xupower@zju.edu.cn. Tel: 86-571-87953226. (Corresponding Author.)

priority of proposed short-distance-bounded-attraction function was demonstrated in complex environment. Simulation results show that the heterogeneous swarm model provides a feasible framework for multi-robot navigation applications.

Keywords: honeybee, heterogeneous swarm, foraging property, short-distance-repulsion

1. Introduction

Animals that travel in groups often rely on interactions among group members to make movement decisions. Social insects are particularly impressive examples, relying on interactions among nest mates they can maintain cohesive behaviors and appropriately respond to environment stimuli. These behaviors have certain advantages such as enhancing the chances of finding food and avoiding predators. In recent years, a variety of efforts have been devoted to modeling and analyzing the swarm behavior, hoping to gain similar advantages on control of multi-agent systems such as multiple mobile robots, autonomous flying machine, etc.

Stephens's paper [1] is one of the early works proposed the foraging theory in animal behavior. In [2], based on the foraging behavior of ant colony, an ant colony optimization algorithm was proposed. Passino[3] and his coworkers used a bacteria inspired model for a swarm moving in an environment with an attraction/repulsion profile. In [4], Gazi and Passino improved their earlier model[3] by adding artificial potential function to the inter-individual interactions and the interactions with the environment, the stability property of the swarm cohesion for different profiles were studied. This model can be viewed as a representation of cohesive social foraging swarms. Almost all the models proposed so far hypothesize that the swarms are homogeneous, in other words, the swarms are composed of the same type of agents with the same functions. For instance, the foraging model in[4] assume that all the individuals use the same dynamic function. Therefore, a disadvantage is that all the individuals need to know the exact relative positions of other individuals and have the ability of sensing the environment. This result in the fact that as the number of the swarm members grows, the computation needed by each agent also grow linearly. In engineering application, such as a multi-robot system based on these models, all the robots need to have excellent detection ability and the production cost will increase accordingly.

Inspired by the organized behaviors of honeybee swarms, an individual-based mathematical model is proposed in this paper for a heterogeneous swarm. The heterogeneous swarm is assumed to consist of two different kinds of individuals, namely, the scouts and the normal agents, with respect to their sensing abilities. Cohesion and foraging properties of the swarm model are studied. In addition, a short-distance-bounded-attraction function is proposed, and simulation results show that this attraction function can significantly improve the foraging accuracy. This swarm model may provide a feasible framework for the implementation of a multi-robot system with heterogeneous sensing capabilities. For instance, a small subset of advanced robots with powerful sensors can guide a large number of simple robots to targets and warn them of potential dangers[14].

2. Biology Evidence

A colony of honey bees can achieve a high level of organization via dynamic division of labor and social interaction. Although there is no leader, the colony is very effective in foraging, comb construction, hive defense, thermoregulation, and other activities[5][11][12]. For example, an intriguing feature of the flight of a honey bee swarm is that only approximately 5% of the bees in swarm have visited the new nest prior to swarm lift off (Seeley et al. 1979). Nevertheless, in the majority of cases a swarm will fly quickly and directly to its destination. Two mechanisms of swarm guidance have been proposed [6]. Lindauer (1955) observed in airborne swarms that some bees fly through the swarm cloud with high speed, seemingly 'pointing' the direction to the new nest site. Lindauer suggested that these fast-flying bees are scouts that have visited the chosen nest site, and that their behavior guides the other, uninformed bees towards their new home. Normally Lindauer's hypothesis is referred as the vision hypothesis. An alternative to the vision hypothesis is the olfaction hypothesis of Avitabile et al. (1975), who proposed that the scouts provide guidance by releasing assembly pheromone from their Nasanov glands on one side of the swarm cloud, thereby creating an odour gradient that can guide the other bees in the swarm. Both the vision and the olfaction hypotheses of swarm guidance seems reasonable, but none of them has been tested empirically until Beekman[7] studied the flights of both normal honeybee swarms and swarms in which each bee's Nasanov gland was sealed shut. The test result proves that only the vision hypothesis is the actual mechanism of swarm guidance.

The clustering in honey bees and in-transit honey bee swarms are spectacular phenomena that have been studied experimentally by biologists during the past several decades. However, many aspects of these phenomena are still not well understood. Here we are only interested in the swarming behavior of the bees after they lift off and the cohesion of the in-transit swarm while moving to the new nest site. Moreover, the inspiration given by honey bee behavior for our research is that in a heterogeneous swarm a small fraction of informed scout agents can successfully guide all the other uninformed normal agents to the destination.

3. Heterogeneous Swarm Model

We consider a heterogeneous swarm including M individuals in an n-dimensional Euclidean space, and model all the individuals as points and ignore their dimensions. Assume that the M individuals can be divided into two different kinds according to their sensing abilities: N ($N<M$) scouts and (M-N) normal individuals. Thus we define the scout rate is $\eta = N / M$. The position of individual i is described by $x^i \in \mathbb{R}^n$. The interactions with environment in this model are based on artificial potential functions, a concept that has been used extensively for robot navigation and control[8][10]. Let $\sigma : \mathbb{R}^n \rightarrow \mathbb{R}$ represent the artificial potential function that model the environment containing obstacles to avoid and targets to move towards. Take the obstacle as a high potential region and the destination as a low potential region. It is assumed that all the swarm members move simultaneously and know the exact relative positions of all the other members, but only the scouts have the ability of detecting the environment. The equation of motion for each individual i can be described by:

$$\begin{cases} \dot{x}^i = -\nabla_{x^i}\sigma\left(x^i\right) + \sum\limits_{j=1,j\neq i}^{M} g\left(x^i - x^j\right), & i = 1,\ldots,N \\ \dot{x}^i = \sum\limits_{j=1,j\neq i}^{M} g\left(x^i - x^j\right), & i = N+1,\ldots,M \end{cases} \tag{1}$$

The term $-\nabla_{x^i}\sigma\left(x^i\right)$ represents that the scout agents can detect the environment and move towards low potential regions (analogous to destination). $g(\cdot)$ represents the function of mutual attraction and repulsion between the individuals. According to the study results on swarming behavior, the effect of function $g(\cdot)$ should be attractive for large distance and repulsive for short distance. In general, the term of attraction/repulsion function that we consider is

$$g(y) = -y\left[g_a\left(\|y\|\right) - g_r\left(\|y\|\right)\right] \tag{2}$$

Where $g_a : \mathbb{R}^+ \to \mathbb{R}^+$ represents the magnitude of the attraction term, whereas $g_r : \mathbb{R}^+ \to \mathbb{R}^+$ represents the magnitude of the repulsion term. $\|y\|$ is the Euclidean norm $\|y\| = \sqrt{y^T y}$. One issue to note here is that the attraction/repulsion function $g(\cdot)$ is odd.

Here we consider a kind of linear-attraction and bounded-repulsion function

$$\begin{cases} g_a\left(\|y\|\right) = a, & a > 0 \\ g_r\left(\|y\|\right)\|y\| \leq b, & b > 0 \end{cases} \tag{3}$$

This function is consistent with the characteristic of the interaction among biological individuals and has been broadly used in the research[4] [9][13]. In this paper we will study the characteristics of heterogeneous swarm using such a function. The attraction/repulsion function can take the form

$$g(y) = -y\left[a - b\exp\left(-\frac{\|y\|^2}{c}\right)\right] \tag{4}$$

Where a, b and c are all positive constants such that $b>a$。

4. Analysis of Swarm Cohesion

In this section, we will analyze the cohesiveness of the swarm and try to find bounds on the ultimate swarm size. To this end, we define the distance between individual i and the swarm center as

$$e^i = x^i - \overline{x} \tag{5}$$

Where $\overline{x} = \frac{1}{M}\sum_{i=1}^{M} x^i$. Therefore, the maximum distance from scouts to the swarm center can be described as

$$e_{s\max} = \max_{i=1,\dots,N}\left\|e_i\right\| \tag{6}$$

Let the Lyapunov function for each individual be

$$V_i = \frac{1}{2}\left\|e^i\right\|^2 \tag{7}$$

Substituting equation (3) and taking the time derivation of V_i, we obtain

$$\dot{V}_i \leq \begin{cases} -aM\left\|e^i\right\|^2 + \sum_{j=1, j\neq i}^{M} g_r\left(\left\|x^i - x^j\right\|\right)\left\|x^i - x^j\right\|\left\|e^i\right\| + \left\|\nabla_{x^i}\sigma\left(x^i\right) - \frac{1}{M}\sum_{j=1}^{N}\nabla_{x^j}\sigma\left(x^j\right)\right\|\left\|e^i\right\|, \\ (i = 1,\ldots,N) \\ -aM\left\|e^i\right\|^2 + \sum_{j=1, j\neq i}^{M} g_r\left(\left\|x^i - x^j\right\|\right)\left\|x^i - x^j\right\|\left\|e^i\right\| + \frac{1}{M}\left\|\sum_{j=1}^{N}\nabla_{x^j}\sigma^T\left(x^j\right)\right\|\left\|e^i\right\|, \\ (i = N+1,\ldots,M) \end{cases}$$

Note that the gradient of almost any realistic profile (e.g., plane and Gaussian profiles) is bounded. Thus, it is reasonable to have the following assumption about the profile.

Assumption 1: There exist a constant $\bar{\sigma} > 0$ such that $\left\|\nabla_y\sigma\left(y\right)\right\| \leq \bar{\sigma}, \forall y$

Then, for $i = 1,\ldots,N$

$$\begin{aligned} &\left\|\nabla_{x^i}\sigma\left(x^i\right) - \frac{1}{M}\sum_{j=1}^{N}\nabla_{x^j}\sigma\left(x^j\right)\right\| = \frac{1}{M}\left\|\left(M-1\right)\nabla_{x^i}\sigma\left(x^i\right) - \sum_{j=1, j\neq i}^{N}\nabla_{x^j}\sigma\left(x^j\right)\right\| \\ &\leq \frac{1}{M}\left[\left(M-1\right)\bar{\sigma} + \left(N-1\right)\bar{\sigma}\right] = \frac{M+N-2}{M}\bar{\sigma} \end{aligned} \tag{8}$$

Since $g_r\left(\left\|y\right\|\right)\left\|y\right\| \leq b$, we have

$$\dot{V}_i \leq \begin{cases} -aM\left\|e^i\right\|\left[\left\|e^i\right\| - \frac{b\left(M-1\right)}{aM} - \frac{M+N-2}{aM^2}\bar{\sigma}\right], i = 1,\ldots,N \\ -aM\left\|e^i\right\|\left[\left\|e^i\right\| - \frac{b\left(M-1\right)}{aM} - \frac{N}{aM^2}\bar{\sigma}\right], i = N+1,\ldots,M \end{cases} \tag{9}$$

Therefore, we conclude the following results:

Lemma 1: Using the linear-attraction and repulsion-bounded function given in (3), consider the heterogeneous swarm described by the model in (1) with scout rate $\eta = N / M$. Assume that the environment satisfies Assumption 1. Then, as $t \to \infty$ the scout's position $x^i(t) \to B_{\varepsilon_1}(\overline{x}(t))$, $i = 1,\ldots,N$, whereas the normal agent's position $x^i(t) \to B_{\varepsilon_2}(\overline{x}(t)), i = N+1,\ldots,M$. Where

$$B_{\varepsilon_k}(\overline{x}(t)) = \{y(t) : \|y(t) - \overline{x}(t)\| \le \varepsilon_k\}, k = 1,2$$

$$\begin{aligned} \varepsilon_1 &= \frac{M-1}{aM}\left[b + \frac{M(1+\eta)-2}{M(M-1)}\overline{\sigma}\right] \\ \varepsilon_2 &= \frac{M-1}{aM}\left[b + \frac{\eta}{(M-1)}\overline{\sigma}\right] \end{aligned} \tag{10}$$

This result is important because it proves the cohesiveness of the heterogeneous swarm and provides an upper bound on the swarm size, which is defined as the radii of the hyperball centered at $\overline{x}(t)$ and containing all the individuals. According to the results above, the ultimate swarm sizes depend on the inter-individual attraction/repulsion parameters (a and b), the parameter of the environment ($\overline{\sigma}$), and etc. Note that the dependence on these parameters makes intuitive sense. Larger attraction (Larger a) leads to a smaller swarm size. In contrast, larger repulsion (larger b) or faster changing landscape (larger $\overline{\sigma}$) leads to a larger swarm size and these are intuitively expected results.

Note that the swarm size decreases with the increasing of individual number M. This is consistent with some biological swarms, where it has been observed that individuals are attracted to larger swarms. However, in biological swarms the number of the members M can be very large and as $M \to \infty$ both ε_1 and ε_2 approach constant values. This implies that for large values of M, the size of the cohesive swarm is relatively independent of the number of the individuals, the scout rate and the characteristics of environment. Note also that while $M \ge 2$ we have $\varepsilon_1 \ge \varepsilon_2$. This implies that, compared with normal individuals, the scouts are farer away from the swarm center. This is because that the scouts can detect the environment, whereas normal agents have limited sensing ability and can only get information about the environment indirectly from the scouts. It means that the link between the environment and scouts is more direct and larger. Therefore, it is much easier for the scouts to be drawn away from the swarm center than the normal agents.

Note also that the scout rate η can only take effect together with environment parameter $\bar{\sigma}$, and larger η (more scouts) leads to a larger swarm size. This is because with larger number of scouts the swarm can obtain more information about environment and thereby the swarm size will be increased. However, decreasing the scout rate η will not have negative effect on swarm cohesiveness. On the contrary it may decrease the swarm size. This result is what we expect and proves the feasibility and rationality of the proposed heterogeneous swarm model.

5. Analysis of Foraging Behavior

In this section we consider the foraging behavior of heterogeneous swarm in realistic profile. The profile we used is a combination of Gaussian profiles. In other words, we consider the profile given by

$$\sigma(y) = -\sum_{i=1}^{K} \frac{A_\sigma^i}{2} \exp\left(-\frac{\left\| y - c_\sigma^i \right\|^2}{l_\sigma^i} \right) + b_\sigma \tag{11}$$

Where $A_\sigma^i \in \mathbb{R}, b_\sigma \in \mathbb{R}, l_\upsilon^i \in \mathbb{R}^+, c_\upsilon^i \in \mathbb{R}^n$ for all $i = 1,\ldots,K$. Note that since the A_σ^i can be positive or negative, there can be both hills and valleys leading to a complex environment.

Note that for this profile Assumption 1 is satisfied with

$$\bar{\sigma} = \sum_{i=1}^{K} \frac{\left| A_\sigma^i \right|}{\sqrt{2 l_\sigma^i}} \exp\left(-\frac{1}{2} \right) \tag{12}$$

Therefore Lemma 1 holds and the swarm cohesion in this profile is proved. Moreover, as $t \to \infty$, we will have

$$\begin{aligned} \varepsilon_{mG1} &= \frac{M-1}{aM} \left[b + \frac{M(1+\eta)-2}{M(M-1)} \sum_{i=1}^{K} \frac{\left| A_\sigma^i \right|}{\sqrt{2 l_\sigma^i}} \exp\left(-\frac{1}{2} \right) \right] \\ \varepsilon_{mG2} &= \frac{M-1}{aM} \left[b + \frac{\eta}{(M-1)} \sum_{i=1}^{K} \frac{\left| A_\sigma^i \right|}{\sqrt{2 l_\sigma^i}} \exp\left(-\frac{1}{2} \right) \right] \end{aligned} \tag{13}$$

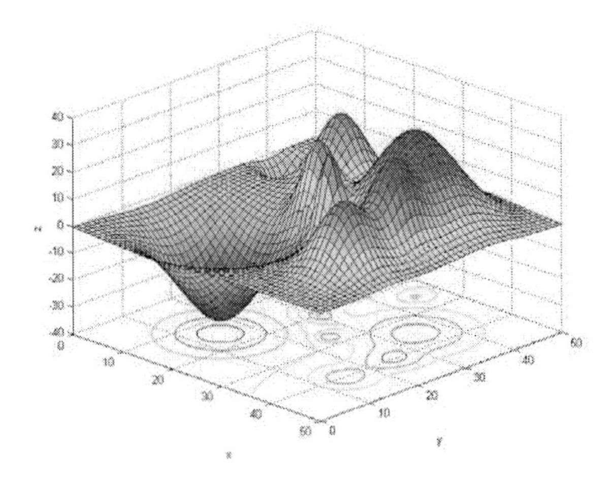

Figure 1. Multimodal Gaussian Profile.

This result specifies the regions in which the scouts and normal agents will converge respectively. In order to study the qualitative properties of swarm foraging behavior, we need to consider the motion of the swarm center. Firstly the distance from the swarm center $\overline{x}$ to the destination point c_σ is defined,

$$e_\sigma = \overline{x} - c_\sigma \tag{14}$$

Using Lyapunov function

$$V_\sigma = \frac{1}{2}\|e_\sigma\|^2 = \frac{1}{2}e_\sigma^T e_\sigma \tag{15}$$

Then, calculate the time derivative of it, in the case of $A_\sigma^k > 0$ we obtain

$$\dot{V}_\sigma^k \leq -\left[\frac{A_\sigma^k}{Ml_\sigma^k}\sum_{i=1}^{N}\exp\left(-\frac{\|x^i - c_\sigma^k\|^2}{l_\sigma^k}\right)\right]\|e_\sigma^k\|$$

$$\times\left[\|e_\sigma^k\| - e_{s\max} - \frac{\sum_{j=1,j\neq k}^{K}\frac{|A_\sigma^j|}{Ml_\sigma^j}\sum_{i=1}^{N}\exp\left(-\frac{\|x^i - c_\sigma^j\|^2}{l_\sigma^j}\right)\|x^i - c_\sigma^j\|}{\frac{A_\sigma^k}{Ml_\sigma^k}\sum_{i=1}^{N}\exp\left(-\frac{\|x^i - c_\sigma^k\|^2}{l_\sigma^k}\right)}\right]$$

Suppose that all the scouts are near c_σ^k and far from other minima. Then we have the following assumption.

Assumption 2: At one minimum c_σ^k, the positions of scouts satisfy the following conditions

$$\begin{cases} \left\|x^i - c_\sigma^k\right\| \le h_k \sqrt{l_\sigma^k},\ h_k > 0, \\ \left\|x^i - c_\sigma^j\right\| \ge h_j \sqrt{l_\sigma^j},\ h_j \ge \sqrt{\frac{1}{2}},\ j = 1,\dots,K,\ j \ne k\ \forall i = 1,\dots,N \\ \frac{A_\sigma^k}{\sqrt{l_\sigma^k}} h_k \exp\left(-h_k^2\right) > \sum_{j=1, j\ne k}^{K} \frac{\left|A_\sigma^j\right|}{\sqrt{l_\sigma^j}} h_j \exp\left(-h_j^2\right) \end{cases} \tag{16}$$

Here h_k, h_j are some positive constants and satisfy the above condition. Therefore

$$\frac{\sum_{j=1, j\ne k}^{K} \frac{\left|A_\sigma^j\right|}{M l_\sigma^j} \sum_{i=1}^{N} \exp\left(-\frac{\left\|x^i - c_\sigma^j\right\|^2}{l_\sigma^j}\right) \left\|x^i - c_\sigma^j\right\|}{\frac{A_\sigma^k}{M l_\sigma^k} \sum_{i=1}^{N} \exp\left(-\frac{\left\|x^i - c_\sigma^k\right\|^2}{l_\sigma^k}\right)} \le \frac{\sum_{j=1, j\ne k}^{K} \frac{\left|A_\sigma^j\right|}{\sqrt{l_\sigma^j}} h_j \exp\left(-h_j^2\right)}{\frac{A_\sigma^k}{l_\sigma^k} \exp\left(-h_k^2\right)}$$

Which implies that if $\left\|e_\sigma^k\right\| > e_{s\max} + \dfrac{\sum_{j=1, j\ne k}^{K} \frac{\left|A_\sigma^j\right|}{\sqrt{l_\sigma^j}} h_j \exp\left(-h_j^2\right)}{\frac{A_\sigma^k}{l_\sigma^k} \exp\left(-h_k^2\right)} \triangleq e_{s\max} + \varepsilon_\sigma$, we can obtain $\dot{V}_\sigma^k < 0$. Therefore, the swarm center will move toward c_σ^k.

Then, consider the heterogeneous swarm with condition in (16) satisfied. As $t \to \infty$, we have $\left\|e_\sigma^k\right\| \le e_{s\max} + \varepsilon_\sigma$. Since $e_{s\max} \le \varepsilon_{mG1}$, combine with the bound implied by (13), we obtain

$$\begin{aligned} \left\|x^i - c_\sigma^k\right\| &\le \left\|e^i\right\| + \left\|e_\sigma^k\right\| \\ &\le \begin{cases} 2\varepsilon_{mG1} + \varepsilon_\sigma, & i = 1,\dots,N \\ \varepsilon_{mG1} + \varepsilon_{mG2} + \varepsilon_\sigma, & i = N+1,\dots M \end{cases} \end{aligned} \tag{17}$$

Thereby, we have the following lemma.

Lemma 2: Consider the heterogeneous swarm described in (1) with the inter-individual attraction/repulsion function as given in (3). Assume that the profile of environment is defined by (11). Moreover, suppose that the locations of scouts satisfy Assumption 2, as $t \to \infty$, we have:

In the case $A_\sigma^k > 0$, for the scouts we have $x^i(t) \to B_{2\varepsilon_{mG1}+\varepsilon_\sigma}(c_\sigma^k), i = 1, ..., N$, for the normal agents we have $x^i(t) \to B_{\varepsilon_{mG1}+\varepsilon_{mG2}+\varepsilon_\sigma}(c_\sigma^k), i = N+1, ..., M$. Where

$$\varepsilon_{mG1} = \frac{M-1}{aM}\left[b + \frac{M(1+\eta)-2}{M(M-1)}\sum_{i=1}^{K}\frac{|A_\sigma^i|}{\sqrt{2l_\sigma^i}}\exp\left(-\frac{1}{2}\right)\right]$$

$$\varepsilon_{mG2} = \frac{M-1}{aM}\left[b + \frac{\eta}{(M-1)}\sum_{i=1}^{K}\frac{|A_\sigma^i|}{\sqrt{2l_\sigma^i}}\exp\left(-\frac{1}{2}\right)\right]$$

$$\varepsilon_\sigma = \frac{\sum_{j=1, j\neq k}^{K}\frac{|A_\sigma^j|}{\sqrt{l_\sigma^j}}h_j\exp\left(-h_j^2\right)}{\frac{A_\sigma^k}{l_\sigma^k}\exp\left(-h_k^2\right)}$$

This result implies that in multimodal Gaussian profile if all the scouts are initially close to one food resource and far from other resources, then all individuals of the heterogeneous swarm will converge to it eventually. Note that the accuracy of foraging depends on swarm cohesiveness and environment condition, and have no direct relation with the scout rate. In other words, the better cohesiveness leads to better foraging accuracy. Thus, with only a small fraction of scouts the swarm may still have good cohesiveness, and acquire high foraging accuracy. This again confirmed the rationality and feasibility of heterogeneous swarm using small number of scouts.

6. Short-Distance-Bounded-Attraction Function

Note that the attraction in (3) has no upper bound and increases linearly with the increasing of y. However, in biological swarms, it is common that the individuals' senses are limited and each individual can sense only the individuals in a limited range. Therefore, the attraction among individuals has an upper bound and will decrease when the distance is too large. In this section, we define a new attraction function called the short-distance-bounded-attraction function to overcome this problem. For the repulsion function we use the same type of function as in the previous section, i.e. functions satisfying (3). The interaction function that we consider is

$$g(y) = -y\left[a\exp\left(-\frac{\|y\|^2}{d}\right) - b\exp\left(-\frac{\|y\|^2}{c}\right)\right] \tag{18}$$

Where *a, b, c, d* are all positive constants such that *b>a, d>c*. Note that *c* and *d* represent the range of repulsion and attraction, respectively. The maximum of Attraction $g_a(\|y\|)\|y\|$ occurs at $\|y\| = \sqrt{\frac{d}{2}}$, and along with the increasing of distance $\|y\|$ the attraction will decrease until near 0. Therefore the attraction represented by this function has an upper bound, and is called short-distance-bounded-attraction. The upper bound and the range of attraction depend on parameter *a, d*. As compared to the linear-attraction function, our short-distance-bounded-attraction function is better consistent with the properties of the interaction among biology individuals, thus is a more reasonable and effective interaction function.

The Fig 2 shows the plot of a short distance-bounded-attraction/bounded-repulsion function (*a*=0.01, *b*=0.4, *c*=1.0, *d*=400) and the plot of a linear-attraction/bounded-repulsion function (*a*=0.01, *b*=0.4, *c*=1.0). As one can see, in short distance (small value of y) the two function plots look quite similar. Note that the attraction of short-distance-bounded-attraction/bounded-repulsion function will decrease while the inter-member distance is larger than 14, and the domain of this attraction is approximately less than 40.

It is easy to see that the short-distance-bounded-attraction and bounded-repulsion functions described in (18) satisfy Assumption 1. However, both the Lemma 1 and Lemma 2 need the attraction function satisfy $g_a(\|y\|) = a$, (*a*>0), and for the short-distance-bounded-attraction function, we can only find an a' in a certain range that satisfy $g_a(\|y\|) \geq a'$. Thus the two lemmas both hold only in this certain range and global convergence cannot be guaranteed. Later we will see that just because of this characteristic, the short-distance-bounded-attraction function can show better result than linear attraction function.

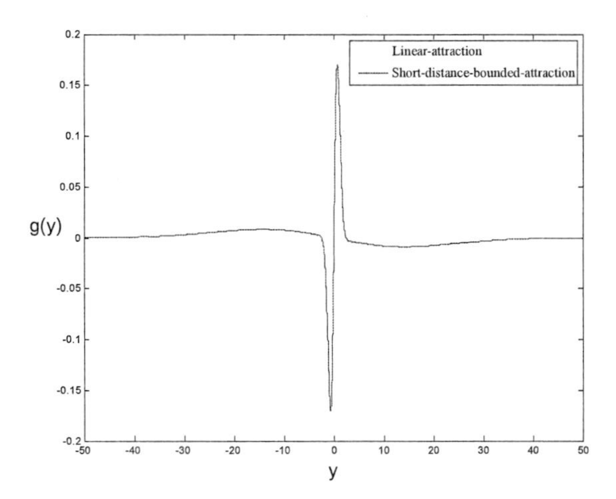

Figure 2. The comparison between short-distance-bounded-attraction function graphic and linear-attraction/bounded-repulsion function graphic

7. Simulation Results

In this section, we will provide some simulation examples to illustrate the theory developed in the preceding sections. The multimodal Gaussian profile we used is shown in Fig.1, which has three minima and six maxima. The global minimum is located at [15, 15], other two local minima are located at [5, 40] and [25, 45] respectively. In all the simulations performed below we chose the swarm with M=20 individuals and N=5 scouts, which means that the scout rate $\eta = 25\%$. As parameters of the linear-attraction and bounded-repulsion function in (2) we choose a=0.01, b=0.4, c=1.0.

The simulation results shown in Fig.3 are for the swarm using linear-attraction and bounded-repulsion function with scouts initially collecting around different positions in the environment. The normal individuals and the scouts are respectively represented by circles and triangles. The upper two plots in Fig 3 show two example runs for which we initialized all the scouts' positions nearby one minimum and far from other minima. For both of the simulations we can see that the entire swarm succeeds to converge to the minimum. Note that for these cases the Assumption 2 in the Lemma 2 is satisfied and the simulation results support the analysis of preceding sections. In addition, we observe that the center of scouts is overlapped with the minimum.

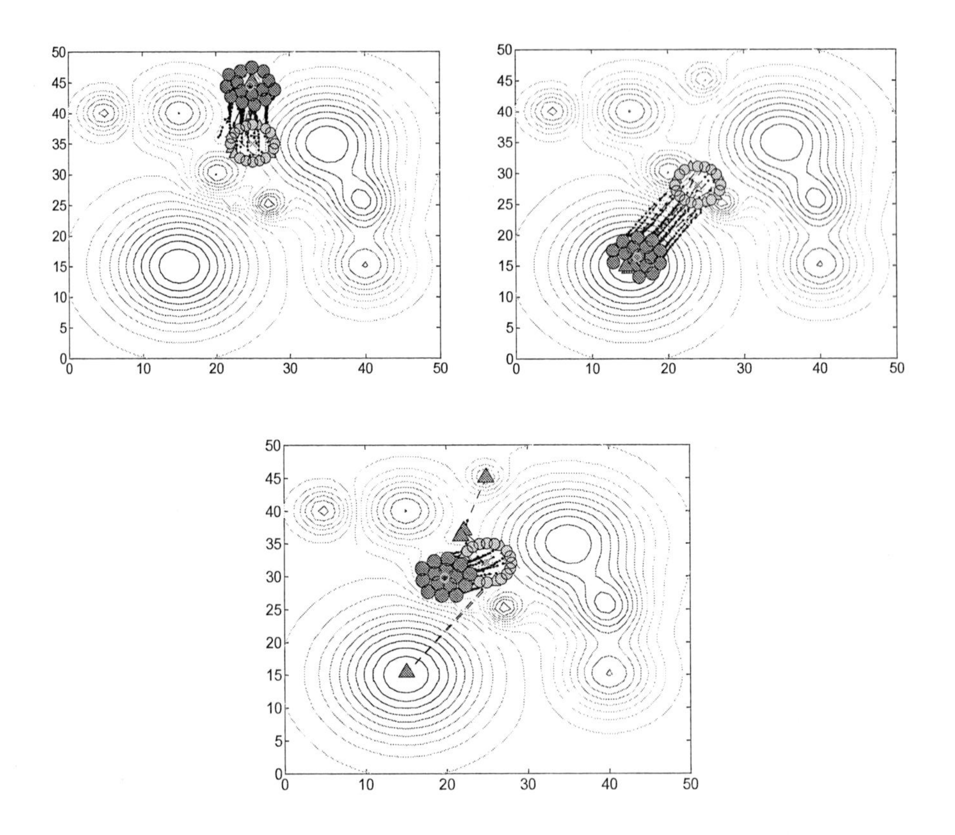

Figure 3. Simulation Results for swarm using linear-attraction and bounded-repulsion function (initial positions of scouts close to a minimum and the counter example).

If Assumption 2 is not satisfied, for instance, some scouts' positions are nearby one minimum while other scouts' positions are nearby another minimum, then the scouts will disperse (As being shown in the lower plot in Fig 3). Note that the normal individuals converge and stay at a position among the scouts (the attraction and repulsion from scouts balance), instead of moving towards any minimum.

Fig 4 show the simulation results with initial positions chosen at random. Note that in this environment the region of the global minimum is the biggest. Therefore, if randomly distribute the swarm individuals in the profile, the probability that in the end the majority of scouts converge at global minimum will be the largest. As being shown in Fig 4, for both of the simulations we can see that three scouts (60% scouts) find the global minimum at last. Therefore for the normal individuals the attraction in the direction of the global minimum is strongest and this results in the fact that the entire swarm will move toward the global minimum.

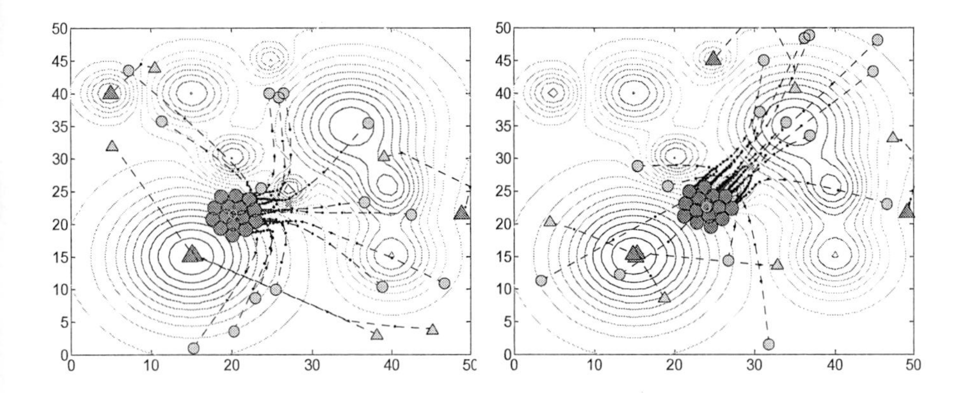

Figure 4. Simulation Results for swarm using linear-attraction and bounded-repulsion function (initial positions of individuals are random distributed)

Therefore, this result shows the advantage of the heterogeneous swarm partly including scouts. In this environment if all the swarm individuals are scouts, a certain percentage of individuals will move to regions far from the global minimum. If the number of individuals in the swarm is large, it will be a big loss. However, in contrast, a small fraction of scouts can successfully direct all the other normal individuals to the destination.

Now consider the interaction function described in (18), as parameters we choose a=0.01, b=0.4, c=1.0, d=400, then the maximum attraction 0.009 occurs at $\|y\| = 14.142$. In the preceding multi-modal Gaussian environment we place two scouts at the global minimum (it means that only 40% scouts can find the accurate food resource position), two scouts at two local minima respectively and the last scout at any other position. The initial positions of normal individuals are chosen at random. Comparing the simulation result of swarm using short-distance-bounded-attraction function (Fig 5 (a)) with the result using linear-attraction function (Fig 5(b)), we observe that the swarm center is much nearer to the global minimum in the first than the second case. Therefore, we can conclude that the simulation result of using short-distance-bounded-attraction function is much better than using linear-attraction function.

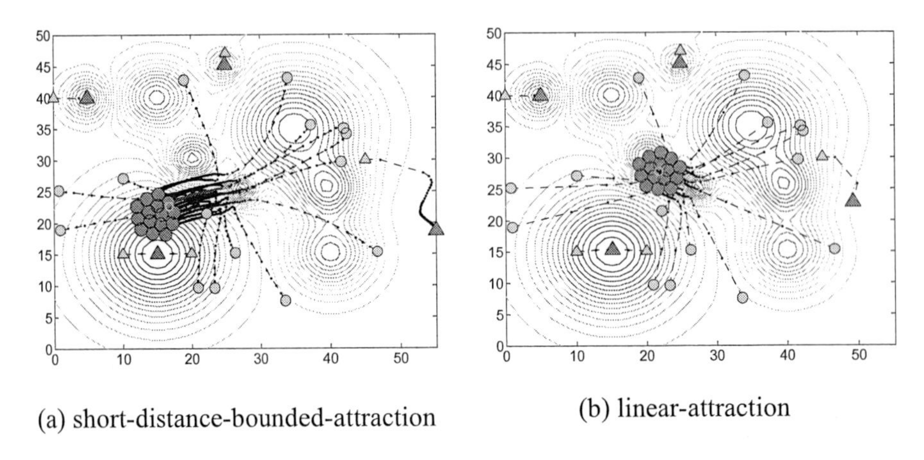

(a) short-distance-bounded-attraction (b) linear-attraction

Figure 5. The comparison of swarm foraging result using different functions.

8. Conclusions

Inspired by the organized behaviors of honeybee swarms, an individual-based mathematical model is proposed in this paper for the heterogeneous swarm. Cohesion and foraging properties of the heterogeneous swarm employs linear-attraction and bounded-repulsion functions in multimodal-Gaussian environment were proved. Furthermore, the priority of the proposed short-distance-bounded-attraction function was demonstrated. The simulation results prove that the heterogeneous swarm can eventually form an aggregation of finite size around swarm center, and converge to advantaged regions of the environment under certain conditions. The results show that a large number of uninformed normal agents can be guided successfully by only a small fraction of informed scouts. Our results might be particularly useful for designing and controlling multi-robot systems and mobile sensor networks, etc.

References

[1] D. W. Stephens and J. R. Krebs, Foraging Theory. Princeton, NJ: Princeton Univ. Press, 1986.

[2] E. Bonabeau, M. Dorigo, and G. Theraulaz, *Swarm Intelligence: From Natural to Artificial Systems*. New York: Oxford Univ. Press, 1999.

[3] K. M. Passino, Biomimicry of bacterial foraging for distributed optimization and control. *IEEE Control Syst. Mag*. vol. 22, pp. 52–67, June 2002.9.

[4] V Gazi, KM Passino, “Stability analysis of social foraging swarms,” *Systems, Man and Cybernetics*, Part B, IEEE Transactions on, 2004, pp. 539–557.

[5] T. D. Seeley. The Wisdom of the Hive: The Social Physiolog Colonies. Harward University Press, Cambridge, Mass, 1995.

[6] S Janson, M Middendorf, M Beekman, “Honeybee swarms: how do scouts guide a swarm of uninformed bees?” *Animal Behaviour*, 2005, 70, 349–358.

[7] M Beekman, RL Fathke, TD Seeley, “How does an informed minority of scouts guide a honey bee swarm as it flies to its new home?” *Animal Behaviour*, 2006, 71, 161–171.

[8] J. H. Reif and H. Wang, “Social potential fields: a distributed behavioral control for autonomous robots,” Robot. *Auton. Syst.*, vol. 27, pp. 171–194, 1999.

[9] Liang Chen, Li Xu, “Collective Behavior of an Anisotropic Swarm Model Based on Unbounded Repulsion in Social Potential Fields,” *8th European Conference*, ECAL 2005, Canterbury, UK, September 5-9, 2005.

[10] Rimon, E., koditschek, D.E.: Exact Robot Navigation Using Artificial Potential Functions. *IEEE Trans. Robot. Automat.* 8(1992) 501 518.

[11] T. D. Seeley and S. C. Buhrman. Group decision making in swarms of honey bees. *Behavioral Ecology and Sociobiology*, **45**:19{31, 1999.

[12] T. D. Seeley, R. A. Morse, and P. K. Visscher. The natural history of the flight of honey bee swarms. *Psyche*, **86**(2-3):103{113, June-September 1979.

[13] V. Gazi and K. M. Passino, “Stability analysis of swarms,” *IEEE Trans. Automat. Contr.*, vol. 48, pp. 692–697, Apr. 2003.

[14] V Kumar, D Rus, S Singh, "Robot and sensor networks for first responders," *Pervasive Computing, IEEE*, vol. 3, issue 4, pp.24—33, 2004.

In: Perspectives in Applied Mathematics ISBN 978-1-61122-796-3
Editor: Jordan I. Campbell, pp. 135-159

Chapter 8

A CAUTIONARY TALE OF CATERPILLARS AND SELECTIONAL INTERFERENCE

W. Garrett Mitchener
College of Charleston, SC, USA

Abstract

Evolutionary theory yields many important insights into why organisms are the way they are. However, for any given problem, there is significant uncertainty about what level of abstraction is appropriate. With such issues in mind, I will discuss an evolutionary scenario concerning the size and development time of a particular species of caterpillar, the tobacco hornworm. A larva of this species grows approximately exponentially; once it reaches a critical size various hormones result in a time delay before pupation during which it continues to grow. Thus, the relationship between genotype and phenotype is non-trivial. In laboratory experiments by Nijhout and in computer simulations, one can selectively breed caterpillars based on size and development time and investigate the resulting evolutionary dynamics. Both laboratory experiments and computer simulations yield striking complexity. Selecting for size and time simultaneously yields unexpected interference, resulting in caterpillars that do not necessarily have the selected-for properties. This interference is due to constraints on the set of possible phenotypes imposed by the caterpillar's fundamental development process. The important lesson here is that when building evolutionary models of even simple phenomena, one must be distrustful of intuition, and consider (1) how the details of the development process of individual organisms affect the ability of a population to explore the space of possible phenotypes, and (2) how selection criteria in combination can interact in unexpected ways.

1. Introduction

The basic premise of biological evolution is that there is a population of individuals that live and reproduce in a more-or-less closed environment. Selection is the process by which the number of offspring an individual has is influenced by its interactions with its environment. Mutation is the imperfection in the reproductive process that causes offspring to be similar but not identical to their parents. These principles are used to understand why organisms have the forms that they do: Presumably, there is (or was) selection in favor of certain features that leads organisms bearing those features to increase in number at the expense of others. The most precise way to reason about selection-mutation processes is to express these assumptions in terms of a formal mathematical or computational model. That precise statement frequently yields a precise solution, and further discussion can focus on whether the assumptions grounding the model are appropriate and how to interpret the solution biologically.

A typical evolutionary model begins by hypothesizing a simplified environment inhabited by organisms playing some sort of abstract game, and the ability of an organism to reproduce depends on its score in this game [11, 23, 24]. Within this general framework, there is plenty of room for variety. The addition of simplifying assumptions often makes the model more mathematically tractable. For instance, if the game can be expressed as a payoff matrix, then mathematical results based on Nash equilibria are available. Alternatively, the model may be more like a simulation, perhaps becoming so detailed that it is mathematically intractable, and the researcher has no alternative but to implement it as a computer program, run it many times, and analyze the results statistically.

Formal evolutionary models can demonstrate complex and unexpected behavior that would go undiscovered if reasoning about evolution were left to imprecise verbal arguments. In particular, models of similar evolutionary scenarios that differ in a few details can have very different outcomes. First, I will summarize two well studied examples that clearly show how details such as spatial structure, environmental limits, and an organism's memory can influence evolution: the prisoner's dilemma and communication games. Both of these focus on the evolution of a single feature of the species in question. Then

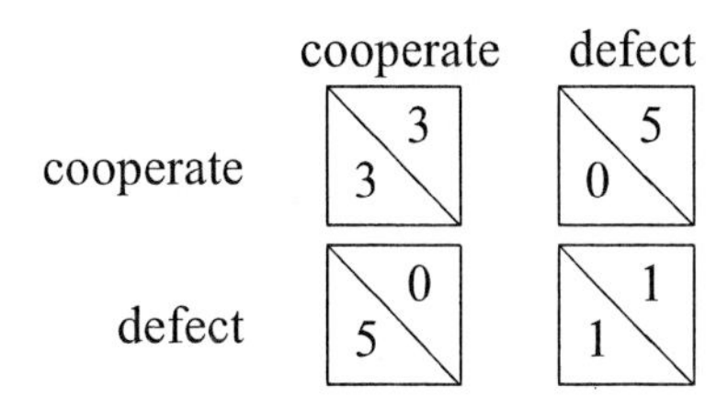

Figure 1. An example of a payoff matrix for one round of the prisoner's dilemma. Each player selects one of the two strategies, *cooperate* or *defect.* One player selects a row, the other selects a column. The combination specifies one of the four boxes. The row player wins the payoff in the lower left corner. The column player wins the payoff in the upper right corner.

I will describe in somewhat more detail some new results concerning a different sort of complexity that occurs when two features of a species are evolving simultaneously: Selective breeding experiments performed by Nijhout's lab on the tobacco hornworm show surprising interactions between selection based on size and selection based on development time [31, 32]. These results are the basis for my simulation that shows how limitations imposed by an organism's basic structure and interference between selectional criteria can prevent a species from developing the traits favored by selection.

1.1. The Prisoner's Dilemma

As a first example, the *prisoner's dilemma* [36] is a family of games in which two players each have the choice of two strategies traditionally called *cooperate* and *defect.* In any given round, both players must choose a strategy without knowing what the other will choose. If they both cooperate, they receive a moderate payoff. If one cooperates and one defects, the defector receives a high payoff and the cooperator is punished severely. If both defect, both receive a low payoff. See Figure 1. This game models the tension between acting individualistically versus cooperatively, and there are countless variations in the mathematical and biological literature. Many of these variations have astonishingly different behavior, given that they are based on the same fundamental game and the same general principles.

In the simplest version of the prisoner's dilemma, the game is limited to a one-round contest between rational players with no memory. Both players should always defect because this maximizes their payoff no matter what their opponent does. In an evolutionary setting of this basic version, there are many players in an unstructured population. Each individual plays the game with many others selected at random, and its payoffs from these interactions accumulate. Individuals with greater total payoff are more likely to be selected to reproduce. In general, the simple strategy ALWAYSDEFECT (which, as its name suggests, always chooses to defect) takes over the population. If an individual attempts to play another strategy, it always scores less than an opponent playing ALWAYSDEFECT, so it will die out.

A more realistic alternative is the iterated prisoner's dilemma. Each pair of individuals plays many rounds of the game consecutively, and their payoffs accumulate. They are allowed to remember the past actions of their opponent during the interaction and take this data into account when deciding on an option for the next round. In this iterated game, players are generally better off risking cooperation. Such strategies include ALWAYSCOOPERATE, which always cooperates; TITFORTAT, in which a player cooperates on the first round and thereafter chooses the action its opponent played in the previous round; and WINSTAYLOSESHIFT, in which a player picks a strategy and uses it repeatedly until it loses, after which it switches to the other strategy, and so on. If the model includes mutation, a population of ALWAYSDEFECT is vulnerable to invasion by mutants that play TITFORTAT because when two such mutants interact they can earn the higher payoff of mutual cooperation, as opposed to the low payoff of mutual defection. So the population tends to change from all ALWAYSDEFECT to all TITFORTAT. However, there is an important effect called neutral evolution. Since ALWAYSCOOPERATE and TITFORTAT are indistinguishable when playing only against each other, a mutation to ALWAYSCOOPERATE is invisible to the selection process in a population of all TITFORTAT. Random drift in the form of such mutations can allow a population of all TITFORTAT players to be replaced by ALWAYSCOOPERATE, which is then vulnerable to invasion by ALWAYSDEFECT [13].

Another variation is to make the population spatially structured rather than well-mixed. Each individual is located at a fixed position on a grid and interacts only with its neighbors. Reproduction is accomplished by cloning an individual into a neighboring grid point, thereby replacing the former occupant. In this case, ALWAYSCOOPERATE can coexist permanently with ALWAYSDEFECT because cooperators can form clusters that enable them to earn the higher payoff of mutual cooperation [36].

In short, the evolution of a population facing something like the prisoner's dilemma depends greatly on details that are not specifically stated in the basic premises of evolutionary theory.

1.2. Communication Games

For a second example that comes from my research, there has been much interest lately in trying to understand the biological origin and genetic history of the language faculty, and in using evolutionary models to understand language change on historical timescales [1–10, 12, 14–16, 18, 20–22, 30, 33–35, 37, 38, 40, 42–45, 48]. One example is the *naming game,* [17, 47] which models a community coming to an agreement on a mapping between words and meanings. Evolutionary game theory determines which such mappings are Nash equilibria or evolutionarily stable. The naming game can be used to precisely describe the insight that when communicating across a noisy channel [39, 41, 46], if there are sufficiently many possible topics of conversation, then the language should switch from using a single symbol per meaning to using a sequence of symbols, that is, a language with syntax.

One can also adapt continuous replicator dynamics [11] to include imperfect learning, with or without genetic variation, resulting in the *language dynamical equation* [19, 25–27, 29]. This equation is a formalization of the idea that language is a strategy for communicating, and the genetically-encoded language acquisition process is a strategy for choosing a language. Typical simplifying assumptions for such models include assuming that the number of relevant languages is finite; that children have a single parent; that they learn from only their parent; that the number of individuals is large enough that a continuous approximation is appropriate, for example by representing the population as a list of what fraction of the population uses each language; that the payoff between players depends on their choices of languages but is otherwise constant; and that each individual interacts equally with all others. Details such as the genetic encoding of language, the syntactic and semantic details of languages, and social structure might be ignored. Mutation might be modeled directly within the resulting differential equations or accounted for as an external perturbation. One case of this model shows how a communication game can lead a population to a coherent state dominated by one language. Another suggests several initially surprising possibilities. For instance, humans may have once been capable of natively learning many more languages than are now possible; that is, some languages may have been removed from the language faculty because selection has favored individuals that do not consider as many possible languages during acquisition and are therefore less likely to make a mistake that inhibits their communication ability [28].

1.3. The Dangers of Excessive Abstraction

The prisoner's dilemma, naming game, and the language dynamical equation all give important insights into the origins and genetic history of cooperation and language, and they lend precision to any discussion concerning them. However, there is reason to be skeptical. Many details, such as the process by which the brain develops and its methods of representing and manipulating meanings are simply not known and are typically left out of models. Furthermore, these models base each simulated individual's reproductive success on its ability to play one abstract game and on nothing else. Evolutionary models in general focus on a single aspect of an organism to the exclusion of all else, which means that interactions among the many demands placed on individuals are left out.

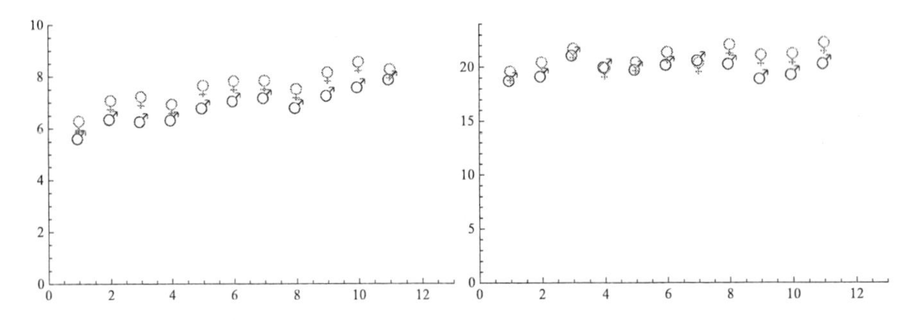

Figure 2. Data from Nijhout's experiments. Left: Mean weight in grams at pupation as a function of generation number. Right: Mean development time in days as a function of generation number. Females: ♀. Males: ♂.

These two issues, the development process and the interaction of selectional demands, are the subject of the rest of this paper. A particular example, the tobacco hornworm, illustrates how dangerous it can be to ignore them. The tobacco hornworm *Manduca sexta* is a well studied species of caterpillar. My interest in the species began with a presentation by Fred Nijhout of Duke University [32], in which he observed that specimens from lab strains were distinctly larger than specimens from the wild. Apparently, something about the

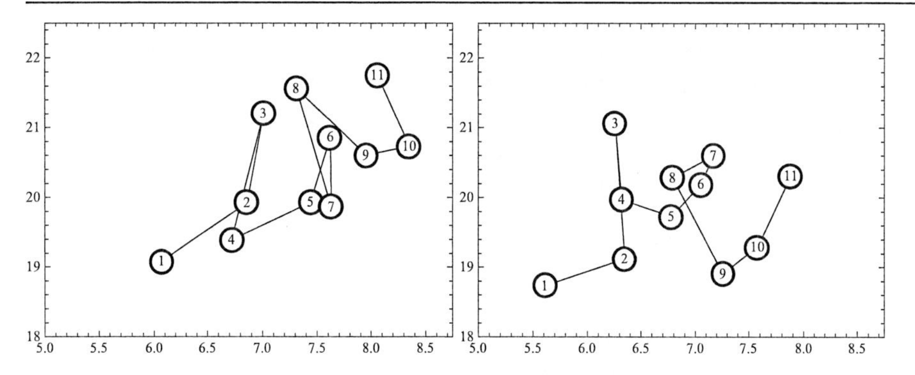

Figure 3. Data from Nijhout's experiments: Mean weight in grams vs. mean time to pupation in days. Numbers indicate the generation. Left: females. Right: males.

lab conditions favored larger caterpillars, and they evolved accordingly. That inspired a series of experiments in which his lab bred colonies of these caterpillars, selecting pupae based on various combinations of mass at pupation and time to pupation.

These experiments had very interesting and unexpected results. One might expect that there should be harmony between development time and pupa size: A caterpillar that stays a caterpillar longer before pupation ought to be larger. Selecting simultaneously for longer development time and larger pupa mass eventually yielded caterpillars with the desired combination of traits, but the path through phenotype space taken by the population was erratic, and frequently backtracked. That is, selecting for longer development time and more massive pupae sometimes yielded faster development times or smaller pupae in the next generation. See Figures 2 and 3 for graphs of some of Nijhout's preliminary data. The populations in these experiments consisted of 221 to 366 caterpillars per generation. Although Figure 2 shows a general upward trend in pupation size and development time, neither feature is increasing monotonically. As Figure 3 shows, the two features also fluctuate independently. Further statistical analysis must be done on this data to determine how significant the trends and fluctuations are, but the overall picture is that selective breeding for longer development time and larger pupae simultaneously, even in a reasonably large population and over 11 generations, does not immediately lead to the expected results. Selecting for shorter development time and smaller pupae also yielded the desired combination of traits only after meandering through phenotype space. Oddly, selecting non-harmonious combinations (shorter time and larger, longer time and smaller) yielded the same general behavior, in which case the eventual success of the selective breeding was the surprising aspect.

Nijhout explained that the hornworm's development process is a complex sequence of hormonal interactions, resulting in a very jagged mapping from biochemical reaction parameters to the development time and final size. Thus, a small change in the genome leading to a small change in some protein could yield a large change in phenotype.

2. A Simulation

In the rest of this chapter, I will describe a computational simulation of Nijhout's caterpillar experiments. It allows the experimenter to directly vary the details of the development process and to adjust the combination of selection pressures placed on the population. The simulation can run on very large populations for many more generations than would be possible in the lab.

2.1. Details of the Computer Program

The computer program simulates the growth and evolution of tobacco hornworms under a variety of selection pressures. It includes some details of the caterpillars' development that turn out to be crucial to understanding Nijhout's results. Each simulated caterpillar is characterized by three genetically specified numbers, k, ρ, and δ. From an initial size of 0.1 at time 0, it grows exponentially at rate k until it reaches ρ times its original size. Living caterpillars are kept in their larval state by the presence of a juvenile hormone, which begins to break down once they reach this critical size. They do not pupate until the juvenile hormone's concentration decreases below a certain threshold. In the simulation, the breakdown of this hormone is represented by a time delay δ during which the simulated caterpillars continue to grow. As a further complication, living caterpillars have a daily hormonal cycle that only allows pupation to begin at a certain time of day. Once the juvenile hormone is gone, the caterpillar must wait (perhaps most of another day) to pupate, during which time it continues to grow. In the simulation, this constraint is represented by requiring that the simulated caterpillars grow beyond the critical size and time delay until time reaches the next whole number.

Mathematically, the time to maturation is

$$t = \left\lceil \frac{\log \rho}{k} + \delta \right\rceil \quad \text{where } \lceil x \rceil = \text{least integer } \geq x \tag{1}$$

and the pupa size is

$$m = m_0 e^{kt} \quad \text{where } m_0 = \text{initial size} \tag{2}$$

To produce the data in this chapter, the simulation maintains a population of 400 virtual insects. Each has a 90 bit genome that encodes three numbers. Each number is encoded by 30 consecutive bits that are interpreted as an integer between 0 and $2^{30} - 1$, then divided by 2^{30} to yield a real number between 0 and 1. These numbers, a, b, and c, determine k, ρ, and δ as follows.

$$\begin{aligned} \rho &= 5 + 100a \\ k &= \frac{\log 2}{0.5 + 4b} \\ \delta &= c \end{aligned} \tag{3}$$

Thus, the critical size ratio varies from 5 to 105, the doubling time for the exponential growth (that is, $\log 2/k$) varies from 0.5 to 4.5 days, and the time delay representing the breakdown of juvenile hormone varies from 0 to 1. (These ranges were selected to give reasonable results and are not explicitly based on any experimentally derived estimates.)

Each generation is constructed from the previous by selecting part of the population for reproduction (details to follow). From the reproductive subset, the simulation picks a pair of genomes uniformly at random, and recombines them at a crossover point selected uniformly at random. After crossover, one of the resulting genomes then undergoes mutation, in which each bit is toggled with probability 0.05. This genome is added to the next generation and the other is discarded. Four hundred repetitions of this process produce the next generation.

In living sexual organisms, crossover occurs during the formation of sperm and eggs cells and involves parallel chromosomes within a single individual. Strictly speaking, the step in the simulation that performs crossover between individuals is biologically unrealistic. It is used here because it enables the simulation to make use of gene recombination without the added complexity of diploid genomes. The program has an option to use simple asexual reproduction instead, but the results are essentially the same either way.

The selection process is very flexible. The four basic criteria are to take the largest or slowest half, or the half with the fastest or slowest development. The simulation can be configured to combine several of these. For example, it can select the fastest half of the largest half, thereby choosing one quarter of the original population for reproduction. Alternatively it can select the largest half of the fastest half, which results in a slightly different quarter of the original population.

2.2. Results of the Simulation

Out of all the possible variations that this program can compute, we will consider the following selection criteria:

- no selection at all
- the fastest half
- the slowest half
- the largest half
- the smallest half
- the fastest half of the largest half
- the largest half of the fastest half
- the slowest half of the largest half
- the largest half of the slowest half
- the fastest half of the smallest half
- the smallest half of the fastest half
- the slowest half of the smallest half
- the smallest half of the slowest half

The composite criteria are similar to those used in breeding experiments on living caterpillars performed by Nijhout's lab. The results of these simulations are shown in an appendix in Figures 6 through 18. The simulated population generally settles to a steady state after ten to twenty generations. The summary diagram in Figure 4 depicts the 100th generation of the simulated population under these selection criteria.

A few striking properties are immediately apparent. First, the four basic (non-composite) selection criteria are clearly effective. One hundred generations are ample time to drive the population to the fastest, slowest, and smallest extremes. In the final population for each of these selection criteria, three quarters of the population is quite close to the extreme of the phenotypic range, with some outliers. The one oddity is selection for the largest. The median settles near the upper extreme of possible sizes, and the larger half of the population is concentrated between sizes 80 and 100, but the population remains very diverse, with around a quarter of the individuals spread out between sizes 30 to 80. This much spread is not present in the results of any of the other basic selection criteria.

Among the composite selection criteria, many of the combinations give reasonable results. However, the combinations involving selection in favor of larger pupae show some unexpected behavior. When selecting for larger size and slower development time, the caterpillars evolve slower development time but do not become much larger than they would be without selection. They are concentrated around size 15, which is nowhere near the maximum of 100 or so that occurs when they are selected for large size alone. To understand this phenomenon, it is useful to plot the set of all possible phenotypes, as in Figure 5. This graph shows that the development process confines the set of phenotypes such that caterpillars of size near 100 and development time near 35 are simply not possible. Most of the population ends up around the upper left tip of the set of possible phenotypes. Intuitively, a caterpillar can only develop slowly if k is near the low end of its range. The growth is exponential, and the additional development time turns out not compensate for the smaller rate constant.

Another unexpected result is the fact that selecting the largest half of the fastest half results in caterpillars that develop quickly and are nearly as large as those that result from selecting based on size alone. This success is possible because faster development requires a large value of k. The additional time spent waiting for the juvenile hormone to degrade and the daily cycle allow them to grow to the maximum possible size.

Oddly, reversing the order of the selection criteria leads to failure. When we select the largest half of the fastest developing half, the caterpillars evolve to be a bit larger than with no selection at all, but the bulk of the population is concentrated around size 22, which is nowhere near the maximum. Referring back to Figure 5, the region of possible phenotypes has very narrow spikes for the fastest and largest caterpillars. Selecting the fastest-developing caterpillars first favors the phenotypes just below these spikes, and the largest of these are relatively small. Selecting the largest caterpillars first favors individuals within the spikes, for which the development time is short anyway. It is therefore not surprising that when selecting short development time first, the caterpillars do evolve a very short development time around 4 at the cost of remaining small, but when selecting larger pupae first, the caterpillars take a little extra time to develop but grow much larger. This particular development process happens to exaggerate the size difference more greatly than the time difference.

If we select for smaller pupae, then as seen in Figure 5, the possible range of development times is much greater, and there is less interference between selection for small size and selection for faster or slower development. Selection for small size and fast development is quite successful. Even though small size and longer development time seem to be contradictory criteria, the combination is within the range of possible phenotypes and is found by the evolutionary process. However, the smallest pupae can only be obtained with development times less than 25 or so because of how the set of possible phenotypes curves away from the vertical axis in Figure 5. Thus, their development time is not nearly as long as it could be without selection for small size.

3. Conclusion

The first lesson to be learned from these experiments is that the devil is in the details. The development of these caterpillars has some complexities that restrict the space of possible phenotypes. Selective breeding can evolve larger or smaller caterpillars, or lengthen or shorten their time to pupation. However, when selection is applied to both characteristics at the same time, there is interference between them, yielding unexpected results. Combinations of traits that at first glance appear to be harmonious, such as larger pupae and slower development, are impossible. Seemingly disharmonious combinations, such as larger pupae and faster development, evolve surprisingly easily. Furthermore, the details of the selection process, such as the order in which the selection criteria are

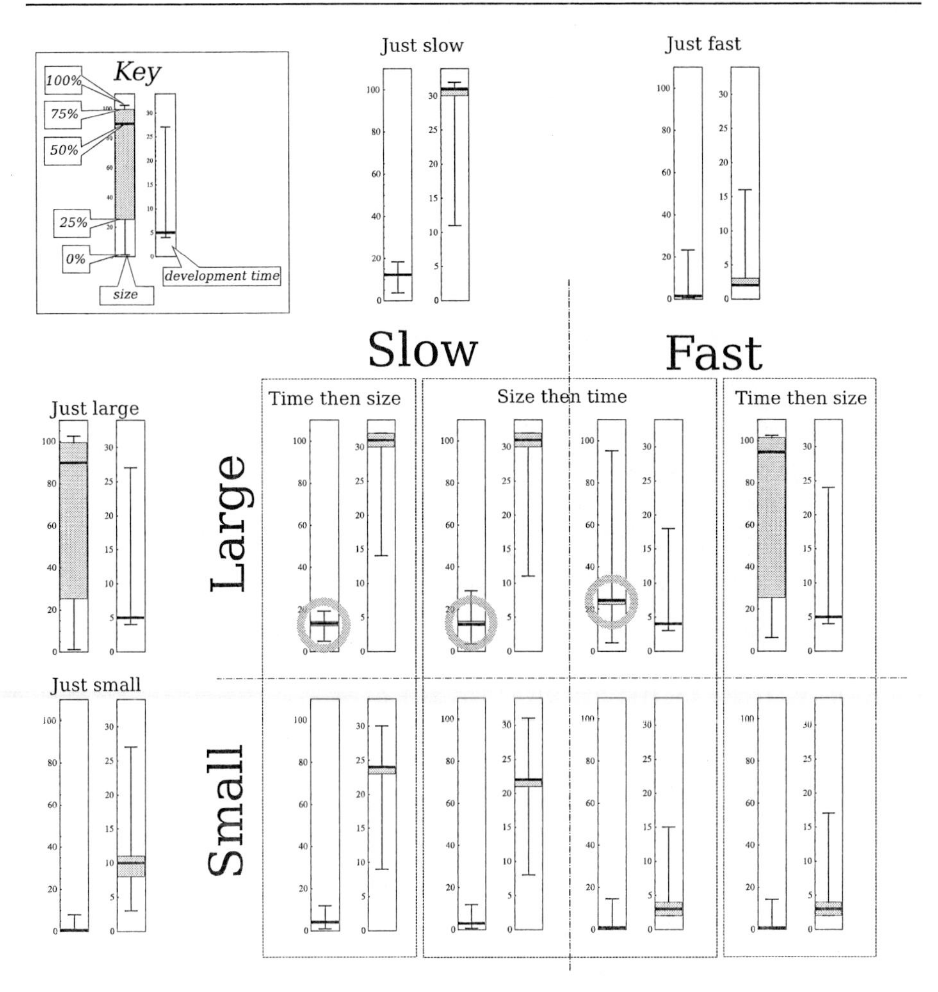

Figure 4. A summary of the results. The box-and-whisker plots for the 100th generation under each selection criterion are shown. Instances where the selection criteria do not lead to the specified results are circled.

applied, are sometimes highly significant.

Genomes in this particular simulation are limited to 90 bits that control three biochemical parameters that lead deterministically to two phenotypic numbers. The living caterpillars that inspired this simulation have much larger genomes and far more potential for variation, so at first it may seem unjustified to claim

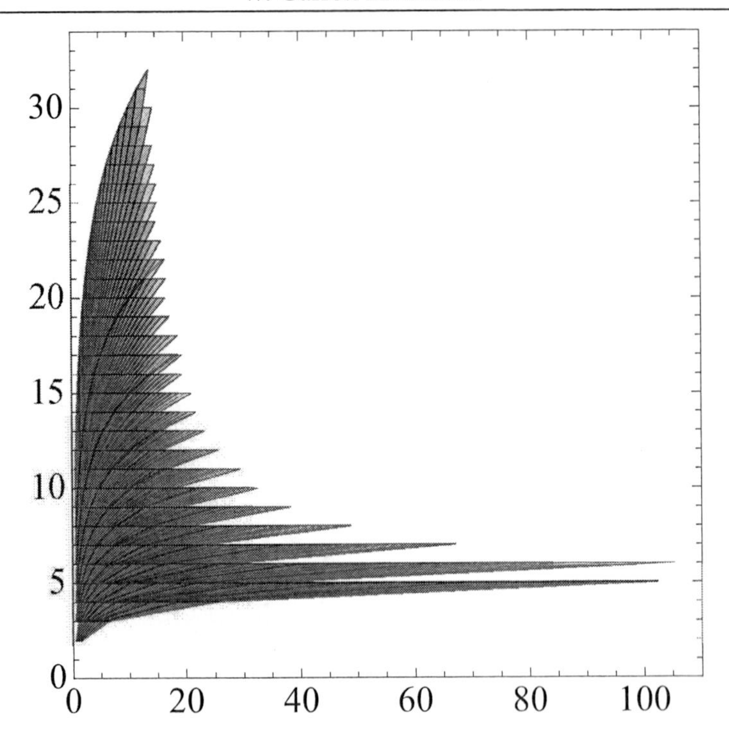

Figure 5. The set of all possible phenotypes. The horizontal axis repesents size. The vertical axis represents development time. This figure was created by plotting the phenotype parametrically as a function of ρ, k, and δ across their respective ranges. The darker areas are phenotypes that happen to be covered by more than one value of (ρ, k, δ).

that certain phenotypes are impossible. However, these caterpillars are very complex biochemical machines. Consequently, many possible variations are fatal. Since the evolutionary path through genome space must always be from one viable genome to another, these fatal variations form barriers that confine the species to a certain region of viable genomes. A long jump from one region to another is always possible since enough simultaneous mutations could happen, but the small probability of such events makes them unlikely to occur within the one hundred generation range of this study. Thus from any initial population and with a fixed time scale, any selection-mutation process will be limited in how much of genome space it can be expected to explore, and it is reasonable to speak of certain genotypes and phenotypes as being impossible. Any species therefore has a limited set of possible phenotypes that it can achieve in the foreseeable future, and the shape of that set can generate the kind of interference between selection criteria see in the caterpillar simulation.

These details about possible phenotypes are typically ignored when setting up a model based on evolutionary game theory. For example, when studying the prisoner's dilemma, it seems at first perfectly reasonable to suppose that tit-for-tat is an available phenotype. However, implementing tit-for-tat in a living social organism requires an episodic memory and a reliable means of identifying each other individual in the community. In mammals, such abilities are common. However, in bacteria or insects, they are almost inconceivable. Common sense does not necessarily correctly identify which phenotypes are possible and which are not. In the caterpillar example, it was genuinely surprising to Nijhout and to me that the combination of slow development and large size is so difficult to achieve, and this fact only makes sense in light of detailed knowledge about the caterpillar's development process.

Evolutionary game theory models also tend to focus on variation in a single trait. In the caterpillar simulation, selection for size or development time alone yields nothing out of the ordinary: The population becomes larger or smaller or slower or faster in agreement with a single selection criterion. Yet selecting for various combinations of these traits results in very interesting outcomes. The simulation suggests that the two traits interfere with each other primarily because the set of possible phenotypes is oddly shaped. Surely this same condition must affect other combinations of selectional pressures in other organisms.

Returning to my own primary line of research, it is clear that these complexities also affect the evolution of language. The growth of a brain capable of symbolic processing from a single cell is at least as complicated as the cartoon of caterpillar growth simulated here. The human communication system clearly grants our species a tremendous survival benefit, but it competes with selection in favor of less powerful but more energy efficient brains, faster growth from a helpless baby, and safer swallowing, for example. The set of possible linguistic phenotypes on the time scale of human evolution is almost completely unknown. The set of all selection criteria and their relative strengths in combination is likewise unknown. In light of the results of the caterpillar experiment, the problem of understanding language evolution seems almost impossible. However, the simulation does provide some suggestions on how to proceed.

First, given some knowledge of the biochemical workings of the tobacco hornworm, the paradoxical results of Nijhout's selective breeding experiments do begin to make sense. Not all biochemical details are immediately relevant, however, the most important ones such as the juvenile hormone are obviously key once their function is identified. In trying to understand evolution of more complex organisms, we must therefore understand as much as possible about the relevant biological machinery if we are to have any hope of correctly identifying the possible phenotypes. When studying evolution of language, for instance, it is imperative for researchers to know as much as possible about how the brain works and how language is learned and used, and to use that knowledge in formulating evolutionary models.

Second, there is a clear need for improved evolutionary models that can handle selectional interference. This does not mean that standard frameworks such as the replicator equation are obsolete, rather, the part of the modeling process in which biological observations are translated into abstract mathematical games must be improved. Instead of assuming that all strategies are possible or focusing on just one or two, a model could hypothesize a restricted continuum of strategies like Figure 4.

Finally, the results of evolutionary simulations and modeling with constrained phenotypes and selectional interference can also aid the understanding of particular organisms. For example, by comparing computer simulations to selective breeding experiments, it might be possible to sketch out the space of possible phenotypes, thereby inferring the actual range of decay rates for variations of hormones.

I would like to thank the NSF for supporting my research (grant number 0734783). I would also like to thank Fred Nijhout and his colleagues for introducing me to this problem and for sharing their data.

A. All Simulation Results

See Figures 6 through 18. In each figure, two graphs are shown, one for pupa size and one for development time. Every tenth generation out of a total of 100 is depicted as a box-and-whisker plot. The heavy line represents the median. The gray box spans the 25th to 75th percentiles, and the whiskers span the 0th to 100th percentiles. Sizes span roughly 0 to 100, and development times span roughly 0 to 35.

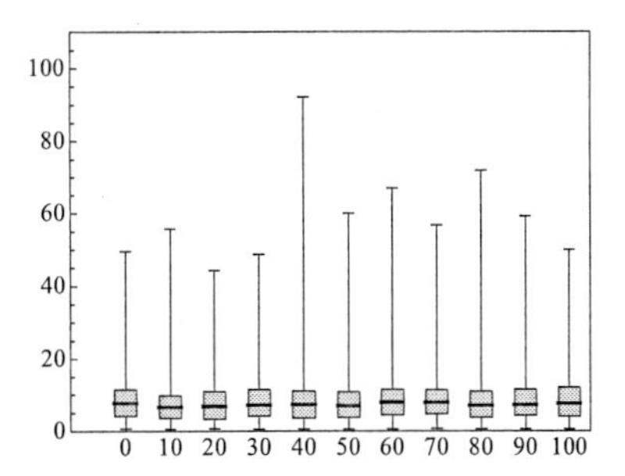

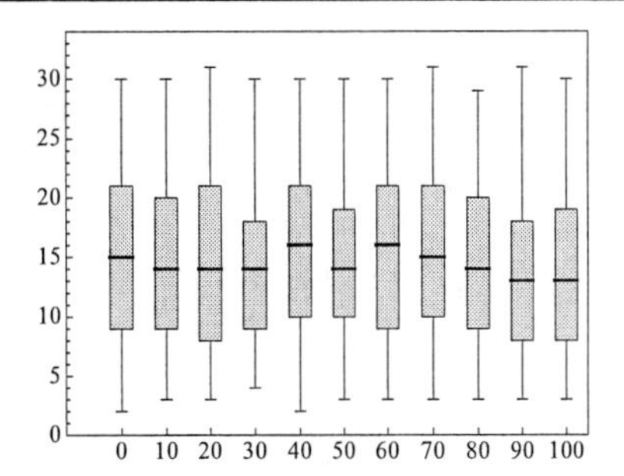

Figure 6. Result of simulation with no selection. Left: Size as a function of time. Right: Development time as a function of time.

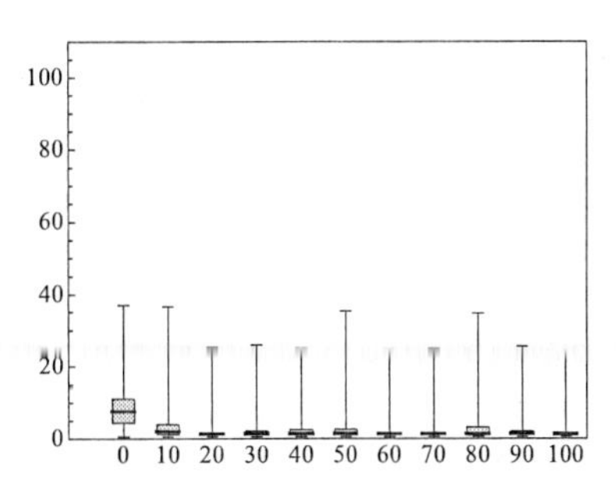

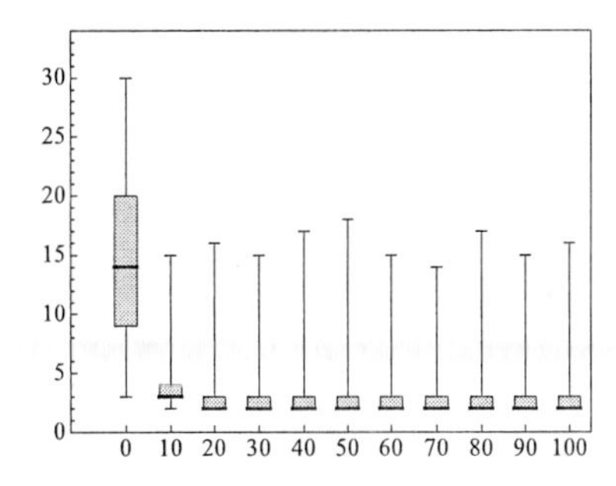

Figure 7. Result of simulation selecting for fast development. Left: Size as a function of time. Right: Development time as a function of time.

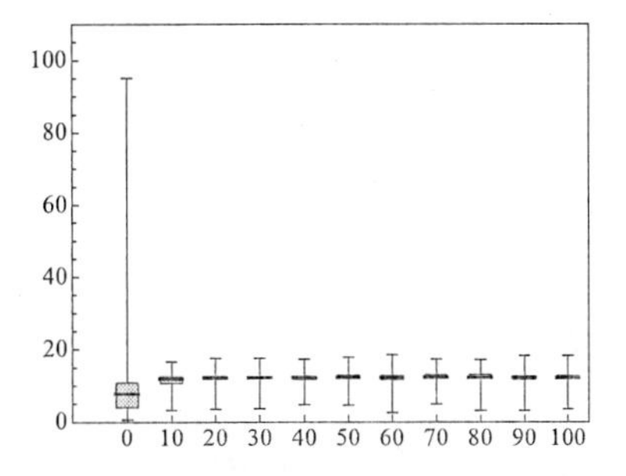

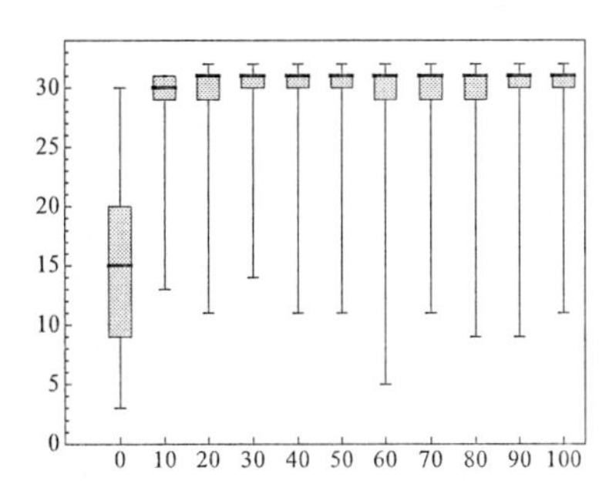

Figure 8. Result of simulation selecting for slow development. Left: Size as a function of time. Right: Development time as a function of time.

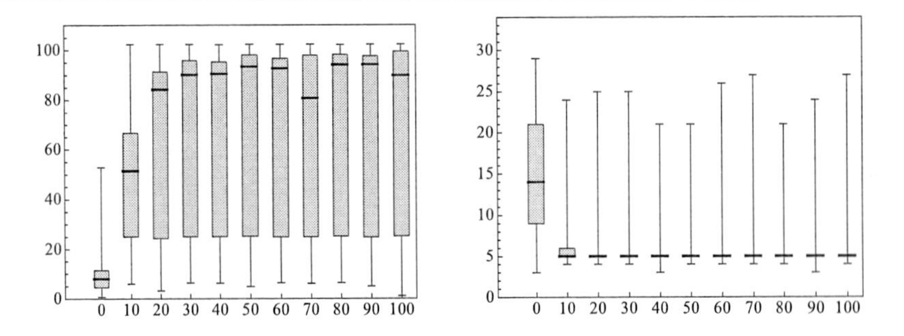

Figure 9. Result of simulation selecting for large pupae. Left: Size as a function of time. Right: Development time as a function of time.

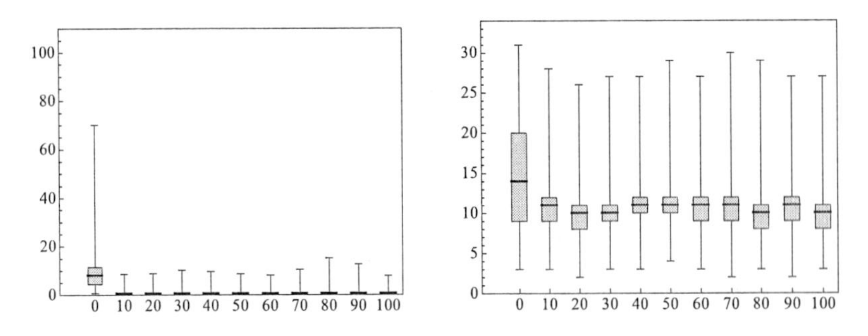

Figure 10. Result of simulation selecting for small pupae. Left: Size as a function of time. Right: Development time as a function of time.

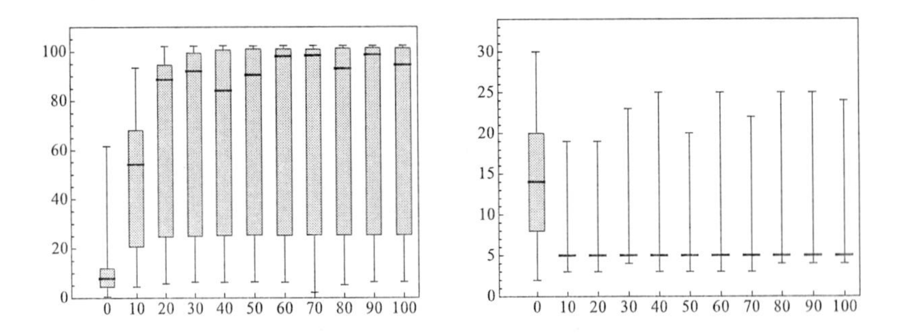

Figure 11. Result of simulation selecting for fast development, then large pupae. Left: Size as a function of time. Right: Development time as a function of time.

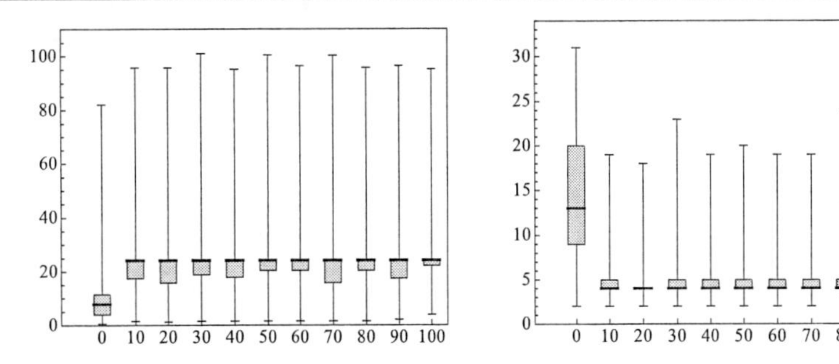

Figure 12. Result of simulation selecting for large pupae, then fast development. Left: Size as a function of time. Right: Development time as a function of time.

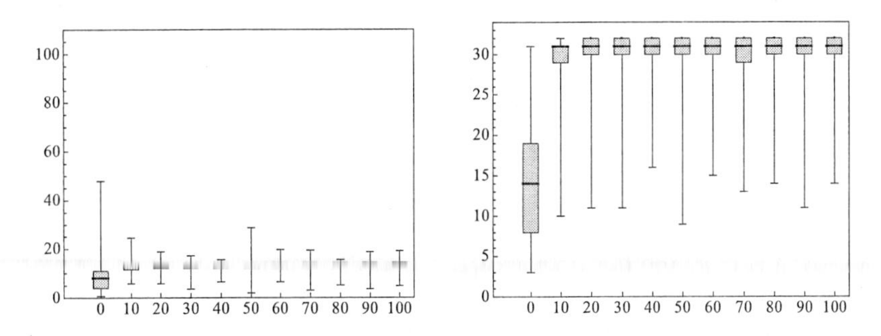

Figure 13. Result of simulation selecting for slow development, then large pupae. Left: Size as a function of time. Right: Development time as a function of time.

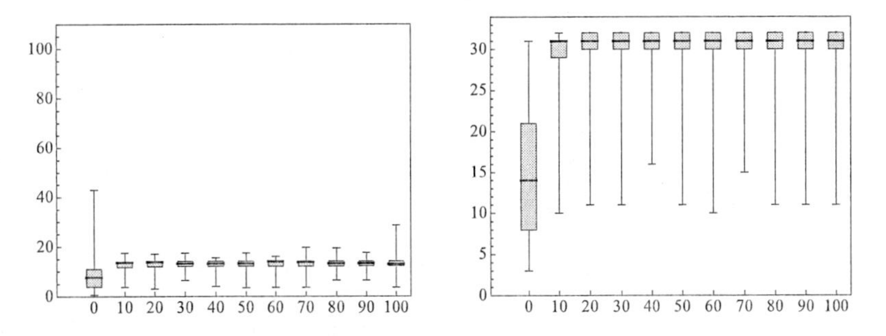

Figure 14. Result of simulation selecting for large pupae, then slow development. Left: Size as a function of time. Right: Development time as a function of time.

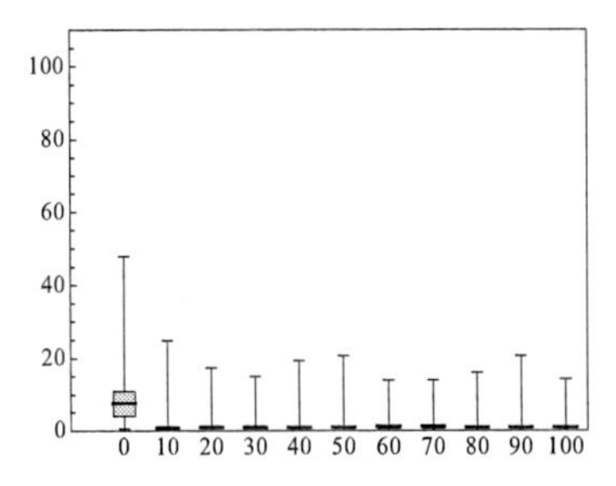

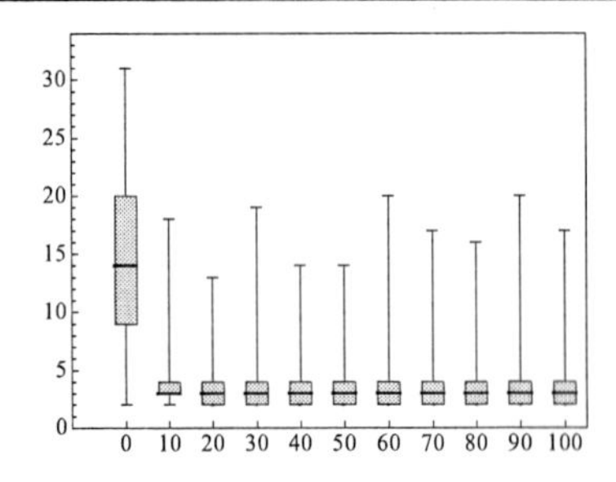

Figure 15. Result of simulation selecting for fast development, then small pupae. Left: Size as a function of time. Right: Development time as a function of time.

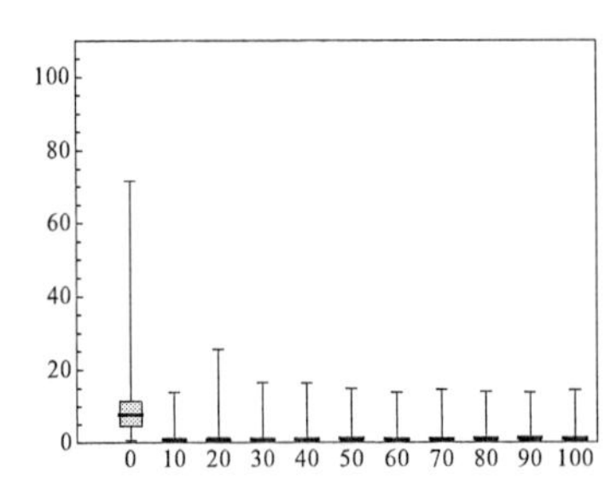

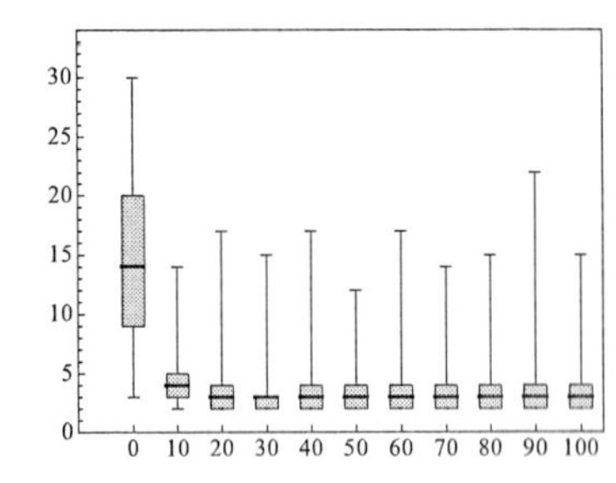

Figure 16. Result of simulation selecting for small pupae, then fast development. Left: Size as a function of time. Right: Development time as a function of time.

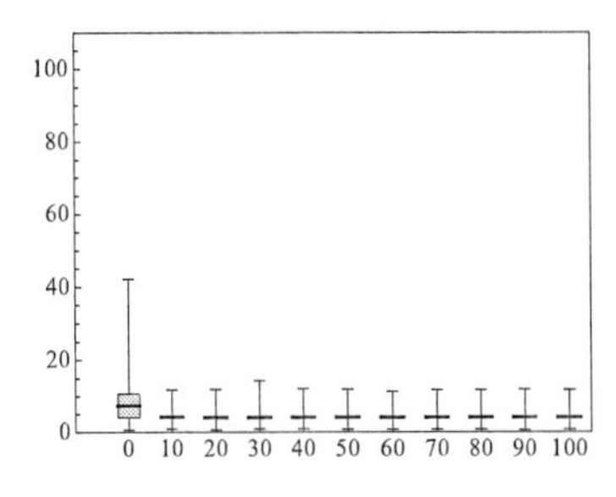

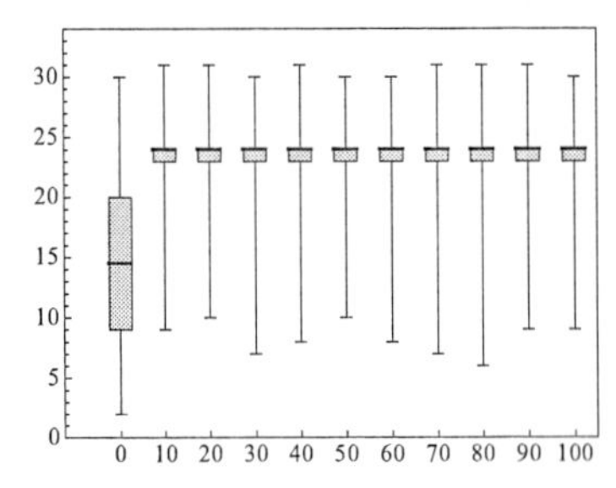

Figure 17. Result of simulation selecting for slow development, then small pupae. Left: Size as a function of time. Right: Development time as a function of time.

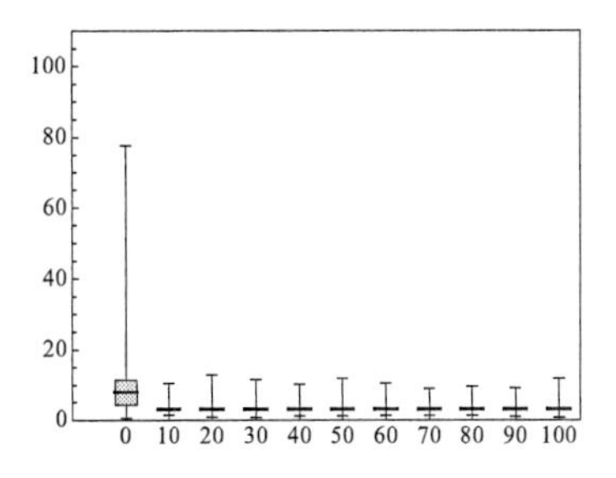

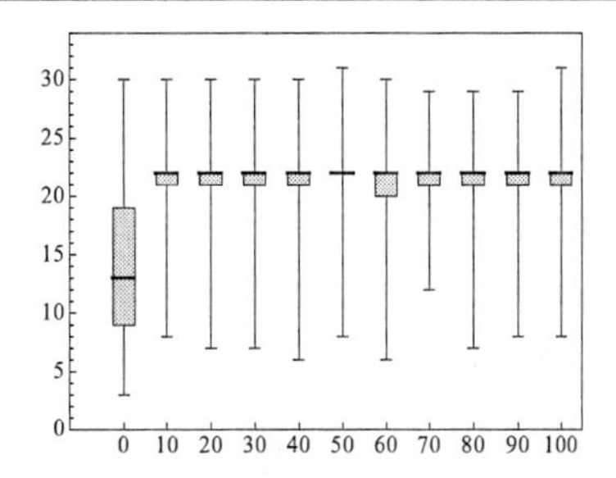

Figure 18. Result of simulation selecting for small pupae, then slow development. Left: Size as a function of time. Right: Development time as a function of time.

References

[1] E. J. Briscoe. Grammatical acquisition: Inductive bias and coevolution of language and the language acquisition device. *Language*, **76**(2):245–296, 2000.

[2] E. J. Briscoe, editor. *Linguistic Evolution through Language Acquisition: Formal and Computational Models*. Cambridge University Press, 2002.

[3] E. J. Briscoe. Grammatical acquisition and linguistic selection. In *Linguistic Evolution through Language Acquisition: Formal and Computational Models* Briscoe [2]. URL `http://www.cl.cam.ac.uk/users/ejb/creo-evol.ps.gz`.

[4] Angelo Cangelosi and Dominico Parisi, editors. *Simulating the Evolution of Language*. Springer-Verlag, 2002.

[5] Wolfgang Enard, Molly Przeworski, Simon E. Fisher, Cecelia S. L. Lai, Victor Wiebe, Takashi Kitano, Anthony P. Monaco, and Svante Pääbo. Molecular evolution of FOXP2, a gene involved in speech and language. *Nature*, **418**(6900):869–872, August 2002.

[6] W. Tecumseh Fitch. The evolution of speech: a comparative review. *Trends in Cognitive Sciences*, 4:258–267, 2000.

[7] M. Gopnik and M. B. Crago. Familial aggregation of a developmental language disorder. *Cognition*, **39**(1):1–50, April 1991.

[8] Marc D. Hauser. *The Evolution of Communication.* Harvard University Press, Cambridge, MA, 1996.

[9] Marc D. Hauser, Noam Chomsky, and W. Tecumseh Fitch. The faculty of language: What is it, who has it, and how did it evolve? *Science*, **298** (5598):1569–1579, November 2002.

[10] John A. Hawkins and Murray Gell-Mann. *The Evolution of Human Languages*. Addison-Wesley, Reading, MA, 1992.

[11] J. Hofbauer and K. Sigmund. *Evolutionary Games and Population Dynamics*. Cambridge University Press, 1998.

[12] James R. Hurford, Michael Studdert-Kennedy, and Chris Knight, editors. *Approaches to the Evolution of Language.* Cambridge University Press, 1998.

[13] Lorens A. Imhof, Drew Fudenberg, and Martin A. Nowak. Tit-for-tat or win-stay, lose-shift? *Journal of Theoretical Biology*, **247**:574–580, 2007.

[14] Ray Jackendoff. Possible stages in the evolution of the language capacity. *Trends in Cognitive Sciences*, **3**:272–279, 1999.

[15] Simon Kirby. Spontaneous evolution of linguistic structure: an iterated learning model of the emergence of regularity and irregularity. *IEEE Transactions on Evolutionary Computation*, **5**(2):102–110, 2001.

[16] Simon Kirby and James R. Hurford. The emergence of structure: An overview of the iterated learning model. In Cangelosi and Parisi [4], chapter 6.

[17] Natalia L. Komarova and Martin A. Nowak. The evolutionary dynamics of the lexical matrix. *Bulletin of Mathematical Biology*, **63**(3):451–485, 2001.

[18] Natalia L. Komarova and Martin A. Nowak. Natural selection of the critical period for language acquisition. *Proceedings of the Royal Society of London, Series B*, **268**:1189–1196, 2001.

[19] Natalia L. Komarova, Partha Niyogi, and Martin A. Nowak. The evolutionary dynamics of grammar acquisition. *Journal of Theoretical Biology*, **209**(1):43–59, 2001.

[20] Cecelia S. L. Lai, Simon E. Fisher, Jane A. Hurst, Faraneh Vargha-Khadem, and Anthony P. Monaco. A forkhead-domain gene is mutated in a severe speech and language disorder. *Nature*, **413**(6855):519–523, October 2001.

[21] Philip Lieberman. *The Biology and Evolution of Language*. Harvard University Press, Cambridge, MA, 1984.

[22] David Lightfoot. *The Development of Language: Acquisition, Changes and Evolution*. Blackwell Publishers, 1999.

[23] John Maynard Smith. *Evolution and the Theory of Games*. Cambridge University Press, 1982.

[24] John Maynard Smith and Eors Szathmary. *The Major Transitions in Evolution*. W. H. Freeman, New York, 1995.

[25] W. Garrett Mitchener. Bifurcation analysis of the fully symmetric language dynamical equation. *Journal of Mathematical Biology*, **46**:265–285, March 2003.

[26] W. Garrett Mitchener. Game dynamics with learning and evolution of universal grammar. *Bulletin of Mathematical Biology*, **69**(3):1093–1118, April 2007. DOI 10.1007/s11538-006-9165-x.

[27] W. Garrett Mitchener. *A Mathematical Model of Human Languages: The interaction of game dynamics and learning processes*. PhD thesis, Princeton University, 2003.

[28] W. Garrett Mitchener and Martin A. Nowak. Competitive exclusion and coexistence of universal grammars. *Bulletin of Mathematical Biology*, **65** (1):67–93, January 2003.

[29] W. Garrett Mitchener and Martin A. Nowak. Chaos and language. *Proceedings of the Royal Society of London, Biological Sciences*, **271**(1540): 701–704, April 2004. DOI 10.1098/rspb.2003.2643.

[30] Frederick J. Newmayer. Functional explanation in linguistics and the origin of language. *Language and Communication*, **11**:3–96, 1991.

[31] H. F. Nijhout. The control of body size in insects. *Developmental Biology*, **261**(1):1–9, September 2003. DOI: 10.1016/S0012-1606(03)00276-8.

[32] H. Frederik Nijhout. The systems biology of body size regulation: A multidimensional mechanism. Presentation at the conference "Applications of Analysis to Mathematical Biology" at Duke University, May 2007.

[33] Partha Niyogi. *The Informational Complexity of Learning*. Kluwer Academic Publishers, Boston, 1998.

[34] Partha Niyogi. *The Computational Nature of Language Learning and Evolution*. MIT Press, Boston, 2006.

[35] Partha Niyogi and Robert C. Berwick. Evolutionary consequences of language learning. *Linguistics and Philosophy*, **20**:697–719, 1997.

[36] Martin A. Nowak. *Evolutionary Dynamics: Exploring the equations of life*. Harvard University Press, 2006.

[37] Martin A. Nowak and Natalia L. Komarova. Towards an evolutionary theory of language. *Trends in Cognitive Sciences*, **5**(7):288–295, July 2001.

[38] Martin A. Nowak and David C. Krakauer. The evolution of language. *Proceedings of the National Academy of Sciences, USA*, **96**:8028–8033, 1999.

[39] Martin A. Nowak, D. C. Krakauer, and A. Dress. An error limit for the evolution of language. *Proceedings of the Royal Society of London, Series B*, **266**:2131–2136, 1999.

[40] Martin A. Nowak, Joshua Plotkin, and David C. Krakauer. The evolutionary language game. *Journal of Theoretical Biology*, **200**:147–162, 1999.

[41] Martin A. Nowak, Joshua Plotkin, and V. A. A. Jansen. Evolution of syntactic communication. *Nature*, **404**(6777):495–498, 2000.

[42] Martin A. Nowak, Natalia L. Komarova, and Partha Niyogi. Evolution of universal grammar. *Science*, **291**(5501):114–118, 2001.

[43] Martin A. Nowak, Natalia L. Komarova, and Partha Niyogi. Computational and evolutionary aspects of language. *Nature*, **417**(6889):611–617, June 2002.

[44] Steven Pinker. *The Language Instinct*. W. Morrow and Company, New York, 1990.

[45] Steven Pinker and Paul Bloom. Natural language and natural selection. *Behavioral and Brain Sciences*, **13**:707–784, 1990.

[46] Joshua Plotkin and Martin A. Nowak. Language evolution and information theory. *Journal of Theoretical Biology*, **205**:147–159, 2000.

[47] Peter E. Trapa and Martin A. Nowak. Nash equilibria for an evolutionary language game. *Journal of Mathematical Biology*, **41**:172–1888, 2000.

[48] Charles D. Yang. *Knowledge and Learning in Natural Language.* Oxford University Press, Oxford, 2002.

INDEX

I

J

K

L

M

S

T

U

V

W

Y